# The Richest Man in Carthage

Catchy Moral Stories with FREE Tools

# The Richest
# Man in
# Carthage

Ammer Mechlaoui

## How to Build Your Empire Using these Bricks and Tools

ISBN: 978-3-00-077467-6

Any references to historical events, real people, or real places are used fictitiously. Names, characters, and places are products of the author's imagination.

The publisher and the author are providing this book and its contents on an "as is" basis and make no representations or warranties of any kind with respect to this book or its contents. The publisher and the author disclaim all such representations and warranties, including but not limited to warranties of healthcare for a particular purpose. In addition, the publisher and the author assume no responsibility for errors, inaccuracies, omissions, or any other inconsistencies herein.

The content of this book is for informational purposes only and is not intended to diagnose, treat, cure, or prevent any condition or disease. You understand that this book is not intended as a substitute for consultation with a licensed practitioner. Please consult with your own physician or healthcare specialist regarding the suggestions and recommendations made in this book. The use of this book implies your acceptance of this disclaimer.

The publisher and the author make no guarantees concerning the level of success you may experience by following the advice and strategies contained in this book, and you accept the risk that results will differ for each individual. The testimonials and examples provided in this book show exceptional results, which may not apply to the average reader, and are not intended to represent or guarantee that you will achieve the same or similar results.

Cover images by the artist @instagram/dalyillustration.
Graphite painting by the Deaf artist @instagram/emna_ghariani_draws.
Book design by the artist @instagram/malek.hajamor.

First printing edition 2024.

Ammer Mechlaoui
Ellesdorferstr. 31
Bonn, Germany, 53179

www.thetyrianpurple.com

# Dedication

Dear readers,

Thank you for buying *The Richest Man in Carthage*!

The content of this book helped me and others to improve their lives.

Now, it is your turn!

I hope you enjoy reading it as much as I enjoyed writing it.

*With much love*
*Ammer Mechlaoui*

# Introduction

*Carthage* was one of the wealthiest cities in the ancient age. To achieve this, their people, the Carthaginians, must have possessed exclusive knowledge. They surely exchanged it only among themselves. It enabled them to dominate the Mediterranean Sea and nearly destroy Rome, one of the world's strongest powers at that time.

The Carthaginian knowledge still exists. It was transmitted over many generations. The place where this transmission happened the most and where this transmission still continues, is Tunisia. The country where the empire of Carthage once existed.

Generally, inherited knowledge is transmitted through stories, idioms and quotes. It is a pleasant way to teach life's lessons. As amusing as this is, this book embeds Tunisian idioms and quotes inside different moral stories. It communicates the essentials for building a wealthy, healthy and balanced life.

## Book Overview

The chapters of this book contain stories. These stories are inspired by the historical events of Carthage. The mentioned traditional Tunisian idioms and quotes are

*highlighted in an italic text.*

1

Each story contains many hidden lessons. After every story, some of the hidden lessons are listed. The reader has the possibility to write his own lessons, too. However, certain lessons need practice. For this purpose, templates are presented as tools.

The next section, *Preferable Historical Knowledge,* contains all the relevant historical information needed to understand our stories. It is recommended to read this information in order to fully enjoy this book.

The richest man in Carthage is mainly written for those who want to find a way to get rich and stay wealthy. It exposes the interrelated beliefs and assumptions of our mindset as the primary reasons for poverty or prosperity. Furthermore, the book introduces the essential aspects for a successful business. It sheds light on different stress situations like having sleep problems or going through hard times. It tackles diverse, relevant issues for a balanced personality like the power of fear and its limitations.

Last but not least, the book considers that the success of individuals is required, but not sufficient, for the success of their nations.

## Preferable Historical Knowledge

The Carthaginian language has its traces in the Tunisian language. For example, *Baal Hammon* was considered to be a weather God[2a]. Until this day *"Baali farming"* means non-irrigated agriculture[2b]. This can be understood as "farming left up to God."

Surprisingly, the Arabic language also uses some Carthaginian terminology. That means Arabs and Carthaginians may have exchanged culture. A part of the first chapter was inspired by this information.

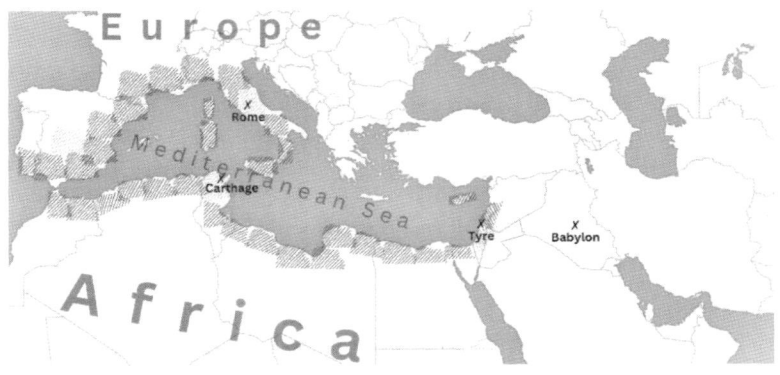

Carthage's territory at its peak.

## Carthage and Babylon

Carthage was found about 300 years before Babylon was disestablished. Knowledge transfer from Babylon to Carthage was therefore plausible. This is relevant for chapter 4. The mother city of Carthage is Tyre.

## Myth of Dido

Dido, also known as Ellisa or Ellisar, fled from Tyre and founded Carthage. A version of her fable is narrated in the beginning of the first chapter.

## Carthaginians

Carthaginians had their own money system that consisted of gold and silver coins. Their most common occupations were sea merchants, farmers, reapers, shepherds.

Carthaginian society was male dominant and prayed to cult stones. They used to sacrifice live animals. Some even pretended that they sacrificed their own newborns. This sounds like madness, but it is proven that Arabs did that. They continued to do so until the Prophet Muhammad deemed it a catastrophic mistake in 600 AD[2b]. All this is mentioned in the first chapter.

## Tyrian purple

The second chapter mentions the Tyrian purple color, a thrilling evidence of the existence of Carthage. It was called so because it came originally from Tyre, but Carthage was famous for it. Tyrian purple was, and still is, extracted from murex shellfish. It is used to dye cloth. Wealthy Carthaginians used to wear Tyrian purple clothes. It became a status symbol.

## Businesses

Small businesses were common:

1. Millers: used windmills to grind cereals.

2. Ceramic makers: formed clay and baked them into ceramic.

3. Stonemasons: worked up marble and chiseled figures, statues, cult stones and much more.

4. Farmers: held animals and planted fields mainly with olive trees and cereals. They hired reapers to harvest cereals.

Big businesses also existed. They were probably the first of their kind. They were often found in the realm of ship production.

## Hanno and Himilco

Hanno and Himilco are known as great navigators. They are the inspiration for the story in chapter 6. They visited faraway lands. The stories of their expeditions were unbelievable for that time. Now, we can understand that Hanno visited South Africa and Himilco might have been the first to reach America.

## Hannibal

A general and statesman who commanded an army of up to 50,000 infantrymen. The army consisted of infantries, cavalry and even saddled elephants. Hannibal was known for his revolutionary strategies. He was also a son and a nephew of high-level officers. His most famous achievement was the battle of Cannae. He caused many losses throughout Roman territory.

## Numidian

They were the native inhabitants of north Africa. They were forced back by Carthage to the western side of the land. One of their kings, Massinissa, caused one of Carthage's major losses.

# Chapter 1 - Dido, aka Elissa(r)

A Carthaginian gold coin (shekel).
Graphite drawing.

# The Believing Mother

Carthage, 300 B.C.

"How can a man bury his newborn daughter alive?"

murmured Sophonisba, on her way to the market with her son, Resheph. They wanted to buy a sheep for the Sacrifice day.

"Why are you saying this?" the little child asked curiously.
"I didn't realize you heard me!" answered the mother, then she continued in a sad voice, "Nowadays, females are less valued than slaves. Do you know that a woman founded this city, Carthage? Her name was Ellisa. She fled from her brother who was hungry for power. When she arrived in this city, she bid the locals to give her a piece of land where she could stay."

"Did they offer her this whole city?" asked Resheph, amazed.
"Locals mocked her. They offered her the leather of an animal that she was sitting on. They allowed her to take as much land as the leather could cover. She was smart and cut the leather into thin strips. These strips surrounded our city, Carthage, that she then ruled. Unfortunately, her brother tracked her. He tried to force her to marry him. She denied and said,

*flame rather than shame.*

She threw herself into fire," narrated Sophonisba.
"Is he the one who is ruling Carthage now, mom?" worried her child.
"That happened before your great-grandfather was born",

8

giggled the mother, "Ages have passed by since then. Nowadays, about three hundred senators rule Carthage. Guess how many women are among them? None! Not a single one, even though female inequality has caused many problems in the past."

"So why are there still no women in the senate nowadays?" wondered Resheph.

"Well-noted, dear son! As a rectification, men engraved a head of a woman onto gold coins. What should an engraving be used for, if a woman still can't even go to the market without a man by her side?"

As soon as they arrived, Resheph was very excited to see the sheep. He quickly ran away. He left before his mother could finish telling the story about burying a newborn daughter alive. Once they arrived, he was fascinated by the animals filling up the place. Some sheep herds were so big that people could hardly walk through them. Resheph chose a sheep with big horns. Although it was too expensive, Sophonisba wanted to please her son. She bought the chosen sheep and went back home together.

Resheph took care of the sheep. He fed the animal and took it for a walk. At the end of the day, the sheep was tied to a tree next to their house. Alone, it began to bleat. Sophonisba worried about thieves, so she woke up several times to check if the sheep was still there.

The next morning, Resheph noticed that his mother looked very tired. She told him that she had had a hard time sleeping because she was worried about the sheep. The little child started to think.

In fact, his mother was not only worried about the sheep, but also about her son. She used to wake up every night to check if he was fine and well-covered.

That brought him to wonder, "Who is more precious? Me or the sheep?"

He spontaneously asked his mother, "What if the thieves took me and forced you to choose between me and the sheep? What would you choose?"

"How did you come up with this question?" asked Sophonisba, astonished.

"You worry about the sheep more than you worry about me", said the jealous child.

"How can you ask that? If I had to choose between you and this city full of sheep, I would always choose you!

*You are my liver! You are the light of my eyes!*"

Sophonisba said confidently.

Resheph remained quiet and didn't say anything. He was imagining the streets of Carthage full of sheep, annoying the people who wanted to walk through. When he realized that he was worth more than them, his self-esteem shot to the sky. His mother interrupted his daydream and said, "Don't be like the majority of men! Be different. Most men desire material things and money the most. Nevertheless,

*money is the dirt of the world.*

They think they have power over money, but in reality, it is vice versa. Even though you are young, don't let money take power over you when you grow up!" The child was worried by these ominous evil powers. He decided to dislike money.

Time passed by and Resheph grew into a unique and confident guy. His mother's lessons had proven their worth, all except for the money lesson. Although he knew that his mother hadn't lied, he couldn't make use of it. Everywhere he met people he found out that they adored money. They would do anything to get it. They said the opposite of what his mother had once said. They believed in the saying,

*"Put money on the dead's mouth, they will smile."*

On the one hand, the majority could not be wrong. On the other hand, his mother could not be dishonest.

During Sophonisba's deathbed confession, she made a heartfelt plea to him, "Don't let the devil mistake you like the rest of the people."
Resheph thought that this might be another superstition like that one about money, but he didn't want to disagree with his mother at such a sensitive moment. He kept her talking.

"It all started when there were no prophets anymore," continued Sophonisba, "There were five men who advised people to do good deeds. When those also passed away, the devil whispered in the ears of humans. And led them to build statues out of stones that resembled these good men. At the beginning, the statue reminded them to do good deeds.
However, over the years, the statues turned and became cult stones. They were considered as Gods. Not only this, when you

11

were a child, Carthaginians met another folk in a faraway desert. These desert dwellers undervalued women and they also prayed to cult stones, like the Carthaginians. Moreover, ashamed by having baby girls, these strangers buried their newborn daughters alive. Both folks exchanged rituals.

As a result, the Carthaginians started to sacrifice their newborn daughters. If they kept doing so, the human race would have disappeared. I believe that whatever happens, the devil will never win. Either the Carthaginians will be destroyed by God's anger, or they will be destroyed by foreign people. I hope that you don't witness that moment, my beloved son. Believe as I do in Jonah and Noah[2d]!"

Resheph didn't want to debate with his mother. He just let her continue talking and just kept listen.

After a few days, Resheph's mother passed away. Even though he was aware that his mother was dying, he was dashed to the ground. She was the most precious thing in his life. Out of despair, he sacrificed a whole camel to a cult stone called Shamash[2e]. He put the offering in front of the statue and begged for his mother to be brought back to life. After praying, he went home and found his mother laying in the same place. He fell apart and started weeping.

It was heartbreaking for Resheph to see his mother so motionless. He just wanted to see his mother's smile for the last time. Suddenly, he remembered the money proverb. He borrowed a piece of gold and put it on her mouth. Needless to say, nothing happened. He fell apart again.

12

Resheph finally recognized that money is just an object. It can't control us. It has no power at all. On the contrary, it's man's greed for possessing money that is the problem. It is not true that

*money turns people evil.*

His mother and other people were not wrong and they were also not right. Money can neither be evil nor revive the dead. He found out that balance is the key.

After burying his mother, Resheph visited the cult stone Shamash. There, he found the rest of all his offerings. All that was left were bones. Disappointed by that, he turned back saying, "Believers in stones are left with bones."


**Some of the hidden lessons:**

*~Mother's lessons~*

- Lessons are not golden rules. Balance is the key.
- In the end, everyone should try to make their own experience.

*~Burying newborn daughters alive~*

- Following the crowd is instinctive, but it can also be fatal.
- What others do must not apply to you.

*~Gold is just an object~*

- Money has no power.

- Money is an object, like a hammer. You can use it for building or for destruction.

*~Your own lessons~*

- …………………………………………..………………………..
- …………………………………………..………………………..
- …………………………………………..………………………..

# Chapter 2 - The Carthaginians

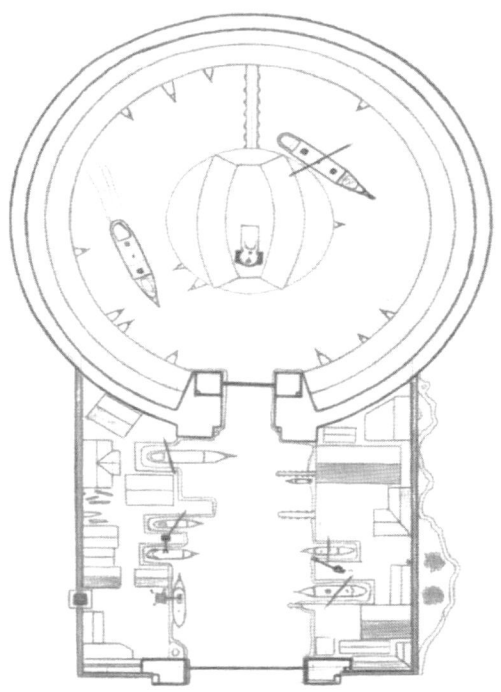

The port of Carthage - Technical drawing.
Graphite drawing.

# Going Even Deeper

The port was the precious gem of Carthage. Its unique form enabled ships to load and unload goods easily. Some of the ships had almost 300 oars. Logically, not everyone could afford large ships. Many small fishermen used little boats with only two oars and a single sail. They usually headed out to sea early in the morning. Catching the same quantity of fish was common. That's why competition between fishermen could not be evaded. It was always a shame to catch less than the rest of the fishermen. Bodo was one of them. He lived alone with his mother in a shack in the surrounding area of the city.

Late one night, Bodo woke up and prepared to go fishing. He tried not to make noise. Despite his efforts, he woke his mother. As he reached the door, she stopped him, saying, "Son! Are you leaving already?"

Bodo looked back and answered, "Sorry mother for disturbing you."

"Oh, no bother!" replied his mother and continued, "I just wanted to make sure that you took your meal with you... and don't forget your hat... and..."

"Mother! Please stop!" Bodo interrupted her and continued, "Don't worry too much. I must hurry up!" The mother felt upset but she couldn't deny that her son was old enough to take care of himself.

*"What is yours is yours and what is not is not,"*

she muttered, on her way back to her bed. It was daybreak when Bodo arrived at the small beach near the big port of Carthage. A few fishermen were already there. He began to clean and repair his net. While he was doing that, an old fisherman addressed him, "You look in a hurry!" Bodo didn't even raise his head. "What's yours is yours!" added the old man with conviction and turned back to work. The fisherman's words caught Bodo's attention. He stopped for a while and looked a bit surprised. "What does that mean? Is it a Goddess sign?" Bodo wondered and kept working, too.

Finally, the net was repaired and prepared and Bodo could depart. Rowing with both oars, he headed to the middle of the bay. Once he reached deep water, he threw the net out in the form of a circle and started to hit the boat with one of the oars. That would scare the fish and ensure that they would be entangled in the net. After a while, the young fisherman started to contract the net.

What a misfortune! Not a single fish had been captured. Meanwhile, another eager fisherman passed by his boat and spoke loudly, "How is it going?! I have gotten the boat half-full of fish. I am already on my way back now!" Disappointed, Bodo rowed to deeper water hoping for a good catch. He threw the net out again and repeated the whole process. All for nothing! This time, a ship passed by and a seaman shouted to him, "A storm is approaching! It's time to return back to the harbor!"

Bodo saw the dark cloud approaching, "I can't turn back with an empty boat!" he thought. Instead of heading home, the eager young fisherman decided to try one last time.

This time he didn't take his eyes off the net while he was collecting it. Just before losing hope, he felt something heavy. Finally, he had been lucky and had caught a big fish. In a hurry, with all his effort, he brought the heavy fish and the net into his boat. Once he started to head back, a strong cold wind blew at him, pushing him away in the opposite direction.

Panicked, Bodo brought down the sail, lest it break. He rowed with all his power but the boat was moving in the other direction. The waves hit the boat harder and harder. The rain started to fall heavily. Bodo covered the boat with the sail. He tried to prevent the boat from being filled with water. Left to the Goddess's mercy, he held the sail as hard as he could. Bodo kept doing this until the waves became less heavy. Eventually, he dared to remove the cover but all that he could see was the sea. Exhausted by the shock and effort, he lost consciousness.

He woke up at night. Oriented by the stars, he spent the whole night rowing. The sun had risen, but there was nothing around except salty blue water. Tirelessly, Bodo kept rowing and this time oriented himself towards the sun. Rendered powerless by rowing and with a thirst to die, he laid down next to his big fish. The bad smell indicated that the fish was already rotten. He definitely needed a miracle.

Miracles can only be made by a Goddess and they need offerings. He thought that only the Goddess could help him now. He could offer the fish, since no one would buy a rotten fish. For humans, it was worth nothing, but for the Goddess, it might be worth his life. Using his remaining power, he threw the rotten fish from the boat. Painfully, he said,

*"Reached the spring but didn't drink."*

Dying of thirst, he lay back and started thinking about his fate. "What if I followed my mother's words and brought food and water with me?" Regrettably, Bodo remembered his mother. "How would my mother be feeling now, and what would she be thinking? Maybe she would be thinking about my last words to her." Thinking about his last words to her - "Please stop" - he said, with a broken heart, "Are these the words that my mother will think about, every time she remembers me?" Bodo started to cry until he fell asleep. The pain of regret had overwhelmed his fear of death.

A short time later, a familiar sound woke him up. It was the sound of a seagull. He wiped his tears away and slowly looked out from his boat. He couldn't believe his eyes when he saw the port nearby. Although he was returning with an empty boat, he had never felt this level of joy before. The other fishermen recognized him and took care of him.

He began to cry as soon as he met his mother again. When he told her the story, she just smiled and said, "Finally you

*saw, what is written on your brow."*

19

Bodo learned that he didn't need to go too far to find the most precious things.

**Some of the hidden lessons:**
*~Fishermen competition~*

- Earning less than others doesn't make the earnings less valuable.

- Earning more than others doesn't make the earnings more valuable.

- Competition is an illusion.

- Never compare your life to another's life.

*~What is yours and what is not~*

- Destiny is unchangeable but it might contain:

- What you want to have is not yours *even if* you do your best,

- What you want to have is yours *only if* you do your best.

*~Going deeper~*

- Trying hard is recommended, but forcing life is not wise.

- Know your limits in order to extend them.

*~The empty boat and the full boat~*

- Happiness starts by appreciating what you have.

- Appreciation needs practice.

- Don't postpone this practice.

*~Your own lessons~*

- ………………………………………..……………………………..

- ………………………………………..……………………………..

- ………………………………………..……………………………..

**Tools for Practice**

The following template helps you appreciate what you already have:

Write what you appreciate most in your life. Cut out the card and use it as a bookmark, for example, or add it to your wallet.

*You can order this template and many other items at*
*www.thetyrianpurple.com*

# Appreciation Cards

Personalise your message, print and cut. Display in your workplace
or leave a surprise note on the recipient's desk and give thanks!

KUDOS!

Estelle,
I appreciate your help this week.
You made a difference!

_____
Olivia.

KUDOS!

Avery
Thanks so much for
taking care of my cat while I was on
vacation

_____
Davis

KUDOS!

Shawn,
Grateful for everything
you do for our family.

_____
Olivia.

KUDOS!

Sebastian,
You did it! You nailed the
presentation and won the client!
Amazing!

_____
Davis

KUDOS!

Carlos,
Congratulations on your
promotion.
So very well deserved.

_____
Olivia.

KUDOS!

Jonathon,
Thank you for your endless care!

_____
Davis

KUDOS!

Claudia,
Thank you for always listening to
me. I couldn't get through it
without you!

_____
Olivia.

KUDOS!

You're the best Pedro!
Thank you for staying positive
during such tough times.

_____
Davis

KUDOS!

KUDOS!

# The Master and Apprentice

It was autumn and the harvest time ended. Adon the reaper lived on the periphery of Carthage. He woke up late and finally enjoyed some time with his pregnant wife and two sons. The first child was named Machaeus and the other boy was called Hasdrubal. Machaeus was fifteen and Hasdrubal was about five springs old. The mother wanted to cook couscous but the cured meat had to be saved for hard times. Buying fresh meat was too expensive. The father decided to teach Machaeus bird hunting. It was a good opportunity to teach him what life disguises.

Adon took some of the wheat from the household provisions. He strewed it about in a spot in the garden, where food waste was usually disposed of. Around the wheat, he installed a net. By pulling the rope, the net would fold in. The father ensured that the net worked fine, by testing it a few times. Ready to hunt, father and son unrolled the rope behind the bushes where they hid themselves.

After a short time, the first bird landed and started to eat. Eagerly, the son said, "Come on! Capture it!"
"This bird is too small. When it keeps eating, it invites more birds to come," the Father answered quietly. The son stared with wide-open eyes.

A short time later, a flock of pigeons landed in the tree above the trap. The son started to count them, "One, two, three... twenty-six birds!"

One bird descended and started to look at the wheat. After a few hesitations, the bird walked into the trap and started to pick some seeds. "Catch it!" screamed Machaeus, failing to be quiet, as he was so excited

"No! We have to wait. One bird is not enough for the whole family," Adon whispered, "These birds are smart. If we scare them, they will not come again, at least for the next few days. Let's keep waiting! The first bird will encourage more birds to come down."

Indeed, a second pigeon came down, as predicted by the father. With less hesitation, the bird walked slowly into the trap. Adon carefully started to prepare for the catch. Machaeus thought of what his father had said before and commented, "No dad, we have to wait, there are still twenty-four birds in the tree!" Adon smiled and answered in a whisper, "Yes son! You have started to learn the basics of hunting." The father's words made his son's pride swell. When a third bird came down directly into the trap, the son proudly said, "Twenty-three birds left!" Without a word, Adon, fully concentrated, waited a few moments but none of the birds followed.

Suddenly, he pulled the rope. The son screamed, "Why! We had twenty-three birds in the tree!"
The father smirked and replied confidently, "I know what I'm doing, the

*touch of a master is better than ten of an amateur,*

what is gone is gone, look what we have got! Three stout birds! Big enough to feed a family of four." Adon knew from his experience that none of the birds from the tree would also come down. He also noticed that their cat was approaching.

Nevertheless, Machaeus kept nagging and didn't show any understanding. Slowly, the smile disappeared from Adon's face. He was a little upset at his son. He decided to teach him a lesson. Without showing any reaction he said, "It's already late son! Let's just go inside."

The mother was already waiting. She was very happy with the hunt. She counted the birds and started to think about how to divide them. Adon took her aside and told her what to do. She was not very convinced, but she just had no time to waste, so she started to cook.

It was evening when the dish was finally ready. The whole family sat around the table. Upon first sight of his plate, Machaeus didn't see any meat. Everyone else had a whole bird on his plate. Quite naturally, Machaeus started to complain. "Father! I hunted with you the whole day, but I didn't get any piece of meat on my plate?"

"Look son! You have more birds than us all together," answered Adon and smirked.
"Where do you see them?" asked Machaeus, cynically.
"In the tree! You have twenty-three birds in the tree!" Adon laughed himself to tears.

The mother tried in vain to prevent herself from smiling. She felt bad for her son and whispered to Adon. She wanted to give Machaeus some of the bird meat from her plate.

Adon refused, "Don't ruin the life lesson that I gave my son today." He continued, "It's not about the meat. The meat comes and goes but this lesson may stay for his whole life. He must always remember,

*three in the hand is better than twenty-three in the tree."*

**Some of the hidden lessons:**

*~Patience is essential for the chase~*

- Chasing wealth is like chasing birds. It is mainly about patience.

*~The small bird invites bigger birds to come~*

- This also applies to money. When small earnings reveal themselves, it is better to invest in them so as to invite big money your way.

*~The touch of the master~*

- Hiring one professional is often better than hiring ten amateurs.

*~Your own lessons~*

- .......................................................................................
- .......................................................................................
- .......................................................................................

# The Shepherds' Undertaking

"There is a treasure under this big tree!"

said Machaeus to his other two younger brothers.
Machaeus, Hasdrubal and Himilco were the sons of Adon, the reaper. During the day, they grazed a few sheep which belonged to Adon's friend. They were not really paid for their service. Their parents wanted them to have an occupation, to learn responsibility.

"Oh please! Don't tell us we have to dig again?!" said Himilco, the youngest child.
"This time I am totally sure!" argued the big brother, "Don't you notice that this tree is the only one in this area? Don't you ask yourself why this rock stone is placed under the tree? It is obviously used to mark the exact location of the treasure. Let's come back at midnight and dig it without being seen by anyone."

*"The greedy spends the night creeping,"*

said Hasdrubal, "I put this rock here days ago to sit on. I tried to hide from the burning sun." Hasdrubal destroyed Machaeus's euphoria.
"We have to change the route. Treasures are hidden in unbeaten roads!" insisted Machaeus.
"We tried all the roads. Since our youngest brother was born. We searched nearly every day, but we didn't find anything. We better give up and try something new," commented Hasdrubal.

30

"Yeah, everything was in vain. By the way, have you heard about Bodo, the fisherman? He discovered mussels in the spot where he fished for a long time. Luckily, these mussels contained big beautiful pearls," added Himilco.

"What a lucky man!" said Machaeus, envying Bodo. Machaeus then complained, "He probably will never need to fish again! It is good for him to own a boat. The sea has a lot of treasures, not like our solid ground here."

"It's amazing to imagine that Bodo floated and passed by his fortune every day!" said Hasdrubal, fascinated. He kept on thinking.

Applying that to him and his brothers could mean that they were probably passing by their fortune every day. Hasdrubal looked around and thought it might even lay beside them at that very moment. They just needed to look closer.

Since that day, Hasdrubal paid attention to what surrounded him most of the time. Sometimes, he thought he might find a spring to sell water. Other times, he thought of catching an extraordinary animal and selling it at a high price. One day, the three brothers heard buzzing. Following the sound, they found bees flying in and out of the ground. They tried painfully to dig out the bee hive and get its honey. Bitterly, they discovered that these were wasps, not bees. They were

*trying to get honey out of wasps.*

Despite this painful experience, Hasdrubal didn't give up. A few days later, he suddenly noticed a piece of wool in a thorny plant. Thanks to his acumen and determination, he realized that these fleeces of wool were widespread. This triggered an excellent idea. First, he and his brothers started to collect the fleeces. At home, they worked up the wool they collected and sold it at the market. Their efforts were crowned with success.

The next day, they left, excited to collect the fleece again. While gathering the wool, they started a new discussion about how they could slowly build their fortune. How they could be as be rich as the lucky fisherman.

"We could buy a chicken then sell its eggs," started Hasdrubal. "If we can make enough money, we could buy two goats and breed them," added Machaeus.
"Then we could buy a dairy cow," affirmed Hasdrubal.
"I want to be the first who taste its milk! I am the youngest," commented Himilco.
"No way! That would be me! I am the oldest," denied Machaeus.

The spoiled Himilco insisted and a fight started. Machaeus beat him up until he bled. Whining, he snitched to his father. Adon was taken aback by what he heard.
He shouted at them furiously, "You? You want to build a business? You can't even be responsible! You are so childish! This will never work.

*Poor partnership leads only to loss.*

I forbid anymore wool collection!"

The children gave up and didn't talk to each other anymore. They lost more than the business. They lost contact.

**Some of the hidden lessons:**

*~Fortune and treasures~*

- Hunting a fortune is much more probable than hunting treasures.

- Fortune-hunting needs to be trained to be seen.

- Fortune may be around at any moment.

*~Honey out of wasps~*

- Painful experiences should not stop you from trying something else.

- Recognize dead-ends before ending dead.

*~Daydreaming~*

- Daydreaming without actions will never make your dreams realized.

- Many worries never happen.

*~The business with the community~*

- Never start a business with someone you don't get along with.

- Businesses can be built out of nothing.

- Don't let the judgment of others destroy your dreams.

*~Your own lessons~*

- ………………………………………..…………………………..
- ………………………………………..…………………………..
- ………………………………………..………………………….

# Chapter 3 - Tyrian Purple

Murex shell.
Digital painting.

# The Rainbow's Seventh Color

"Father, can you tell me the rainbow fairytale again?"

asked Resheph, laying on his bed.

"Again? We've had that yesterday! It is too long!" replied the father, Carthalo.

"Please daddy, I will not interrupt you this time!" begged the child.

"Alright," Carthalo answered, "Let's do it all again.

Once upon a time, the rainbow thought to share its gorgeous seven colors (red, orange, yellow, green, blue, indigo and purple) with humans. First, the rainbow began sharing its most amazing color, red. As it happened, humans were crazy about red. They started to kill each other just to possess it. As a punishment, the rainbow prevented them from having access to this color.

Humans apologized to try and get the red color back. Although the rainbow forgave, it cursed them anyway. It turned the color of their blood red to remind them of their awful deeds.

The rainbow believed in humans and gave them another chance. This time, the rainbow offered them two colors at once. Orange and yellow became the most-beloved colors of humans. They began to overuse both colors.

The rainbow was angry and dried the land. Trees and plants became orange and yellow. Soon, the humans didn't have enough to eat so they started to beg the rainbow. Out of mercy, it forgave them again. This time the rainbow played a different trick on the humans. It spread its green color over the lands and dissolved its blue and indigo colors into the seas. It kept purple, its most precious color. It not only hid the last color in deep water, under rocks and waves, but also scattered it within hard spinous armors. No one was able to get at it anymore.

Ages passed by and the rainbow became too old. It was afraid that the purple color would fall into the wrong hands. It chose an innocent, wise and brave man to reveal the place of the purple color. The chosen man kept the secret. In return, the rainbow decided to reward him. It gave him a recipe describing how to extract purple. The man was allowed to reveal the secret only to his descendants. But first, they must prove that they are worthy!

Since that day, the rainbow disappeared. Its ghost might only appear when the sun shines and the rain falls at the same time.
The chosen man was your great-great-great grandfather. This is how I inherited this secret recipe that I will pass onto you one day." The father finished his tale and found out that the child was already asleep. He left saying,

*"99 ants entered the ant hill."*

One day, as Resheph woke up, he found his father still at home. Resheph was very happy and said, "You are still here! You didn't go to work."
"Yes! I am waiting to take you with me. Today, I will reveal the secret recipe to you!" answered Carthalo.

They took some equipment and left home together. The mother, Sophonisba, kept her eyes on them until she was no longer able

37

to see them. She was worried about her son's first apprentice day. Father and son walked to the cliffs near their house. Both walked down the rocks until they reached the seawater.

"Dear son! Now I will show you where the rainbow buried his purple color. However, in order to do that, you first have to prove that you are worthy of it!"
"I'm worthy of it!" answered the keen son.
"Alright! You have to take a deep breath and dive to the bottom. There, you will find shellfish. Your mission is to bring me one."
"Is that all? Let me show you how I can do that easily, right now!" said Resheph and jumped head first.

After a short moment, he came back to the surface, aching.
"Ouch! My ears are hurting so hard! I thought it was easy to do," the child screamed.
"Good try my boy! It is not difficult, you just need to practice that every day. Whatever happens, just don't try to force it! If you do so, the rainbow may punish you."
"Did he punish you? Can you do that? Is this your craft?" asked Resheph, questioning his father while holding his head between his hands and pressing his ears.
"Ha-ha, bear with me. Firstly, the rainbow didn't punish me because I didn't force it. Second, of course I can do that, it's part of my labor." The father added, "Not just that. I take the shellfish and work them up. I extract the purple color out of them until it becomes a powder. The purple powder is more valuable than gold."

"Can you show me how you do that?" wondered Resheph, excited.

His father gently denied him, "You must prove to the rainbow that you are brave and patient first." He finished by saying, "Enough for today. Let's go home!"

Father and son left to go back home. In fact, it was not an easy task, but the child was excited. He decided to try it until he succeeded.

Days passed by and Resheph was still fascinated by the challenge. He kept practicing every day until he could finally do it. He took the shellfish and ran to his father. Carthalo was very proud of his son and they celebrated this victory.

One day, he was playing 'Chaos'[1a] with his friend, Gisgo. Resheph told Gisgo about his success. The friend was excited to try and wanted to see if he too would be chosen. Both kids went to the sea, to the same cliff where Resheph's father took him before. The friend tried to dive, but the pressure of the deep water prevented him from reaching the bottom. Resheph told him that he also reached the same depth on his first try. Gisgo was jealous, so he quit, saying, "You believe in childish myths. You believe in lies! I don't even need this. My father can simply buy anything I want."

Both kids returned home. Resheph told his father what happened. The father soothed him, saying, "He is jealous and is trying to make you jealous, too. Some people love competitions, like your friend, and others enjoy challenges. You should know what type you are. Competitors would be overwhelmed by the challenger's achievement. Challengers would be bothered by the competitors. Do you know what type you are?"

The child thought for a while and said, "Hum, at the beginning, I really enjoyed the challenge. But the competition with Gisgo frustrated me."

Years passed by, Resheph helped his father to extract and sell the purple color in his shop. The purple powder became more precious. People crossed the sea to buy just a tiny quantity of it. The purple color became a symbol of nobility and the upper class. Only those who were wealthy dressed in purple.

Over the years, both Resheph's and Gisgo's fathers passed away. Resheph mastered his father's labor and inherited his business. One day, a beggar came to Resheph's shop. After talking, it came out that the beggar was Gisgo, Resheph's childhood friend. Although Gisgo inherited the wealth of his father, he had spent it all. After his bankruptcy, Gisgo badly wanted to learn Resheph's labor.Looking at how motivated he was, Resheph decided to give Gisgo a try.

They went to the cliff and jumped into the water, Resheph showing Gisgo what to do. Thanks to many years of experience, it looked very easy. Resheph emerged from the water with a shell in his hand. Gisgo tried to do the same but failed due to the water's pressure. Being ego-driven, he dived again, this time even deeper. Despite all his effort, Gisgo didn't make it. He came up to the surface with a bleeding ear. One of his ears turned deaf. Resheph looked at him with regret and said,

*"Heritage goes away and handcraft will stay."*

40

**Some of the hidden lessons:**

*~99 ants~*

- It is better to make an effort only for those who appreciate it.

*~Seventh color~*

- Protect your business ideas and products like the rainbow's seventh color.

- Choose carefully who you want to share with.

*~Competitors and challengers~*

- It's important to know your type: Competitor or Challenger.

- Working with the same types increases results.

- Working with the different types decreases results.

*~Earnings and heritage~*

- Money earned fast is often hard to keep.

- Money earned more slowly is often easier to keep.

- Training and experience make hard work look easy.

- Money runs out while knowledge stays.

*~Your own lessons~*

- ……………………………………………..……………………………………..
- ……………………………………………..……………………………………..
- ……………………………………………..……………………………………..

# The Instant Pleasure

"Daddy! Why are we poor?"

the half-naked dirty boy asked his father, Gisgo.
The boy interrupted their search in the waste, after the market had closed. They were looking for anything edible such as abandoned fruits and vegetables. It didn't matter if the remains were stale or damaged by transport or parasites.

The father looked embarrassed at his son Rhodian and asked, "How did you come up with this question?"
"I met a boy my age today, on my way here," the boy said, "He had clean clothes and a good smell. We played for a while. I asked him to join our waste search. His mother shouted at him and said that only poor people do that. Is that true?"

The father felt uncomfortable, but answered, "You embarrass me, son. This is a part of my life that I feel really ashamed of."
"Why should you be embarrassed?" wondered Rhodian, "You told me to accept what I am."
"Alright, I will be honest with you. The reason is instant pleasures." The father had finally revealed the truth.
"You mean females and fermented cereal juice?" inquired the child, curiously.
"How do you know about those things?" asked Gisgo, astonished.
"I heard that our neighbor has those problems," said the child, defending himself.

"Impetus is a curse!" confirmed Gisgo then continued, "Some are driven by females and drink. Others are obsessed by games and gambling. My curse was instant pleasure. I could never resist offers from merchants. Every time I see something, I wish to possess it or at least have something similar. Your grandfather never left me in need of something. He was a wealthy man." He paused then said, "Actually, we didn't get along well with each other. He was always busy with his business. Except during his last days when he was sick. He expressed for the first time how much he loved me and that he did everything only for me. He was powerless when he stated his last words. I still remember them very well,

*"Get hungry and enjoy honey,*
*bind your wife before starting a quarrel, and*
*wear then tear healthy."*

"That sounds weird," replied Rhodian.
"That's true, but I felt indebted to him. I was obliged to follow his advice. So, every time I got hungry, I mixed my meal with honey. I tasted different types of honey from all over the world! I ripped my clothes apart after each time I wore them. In three springs, I was married three times. Every time I quarreled with my wife, I bound her to the bed. That's why they all left me immediately.

Every time I ran out of money, I get some more from the moneylender. He knew my father well. When he recognized that I was spending too much, he advised me to pay attention to my expenses. He said if I were to keep this up, I would go broke. Every time he warned me, I ignored him, saying,

*revive me today then kill me tomorrow.*

This kept happening until I ran out of money, once for all."
"Seriously! If that's what grandfather really meant, did he live that way too?"
"Surely not, I have discovered the hidden meaning from the moneylender. When I wanted to borrow money, he asked me for what purpose. I told him my father's last advice. He laughed at me and revealed the meaning:

First, wait until you get hungry and everything you eat will taste as sweet as honey. Second, bind your wife with a strong relationship and love before going through life's troubles. The final one means that, as long as your belongings are not broken and not damaging your well-being, you don't need to renew them.

I thought I was following my father's advice but in reality I was following my own instant pleasure."
Both kept silent for a while. The father was concerned that his son would blame him for their misery. However, Rhodian just needed to digest this new reality. Gisgo broke the silence, "It seems it is our destiny to live poor."

"Why are you saying that?" wondered the son.
"Your grandfather told me that he was like us now before he made his fortune. Allegedly, he met someone that showed him the way out of misery," his father answered, restarting his search in the waste.

All of a sudden, Rhodian had a brilliant idea. He stopped searching and ran away. "Where are you going!?" yelled his father.

"Searching for someone who can show us the way out of misery! Like our grandfather did before!" answered the son without looking back.

The father stood and stared, then said to himself, "Why didn't I think of that before? Probably I'm still blinded by the instant pleasure. I spend every day worrying about how to survive instead of making my life better." This new perspective led him to stop searching in the waste and pursue his son.

**Some of the hidden lessons:**

*~Last few words~*

- Less is sometimes more.

- Necessary things are the most enjoyable ones.

- Build a strong relationship with a woman before going through hard times with her.

- Keep your belongings as long as they are not broken or harmful to your health.

*~Instant pleasure~*

- Controlling your expenses is the key to building wealth.

- Consider tomorrow as a new day to improve your life, not as another day to survive.

- Improve your financial situation by seeking knowledge from rich people.

*~Your own lessons~*

- ……………………………………………..………………………..

- ……………………………………………..………………………..

- ……………………………………………..………………………..


**Tools for Practice**

The following template will help you recognize instant pleasure: Use the maze to check whether you should buy something or not.

*You can order this template and many other items at*
*www.thetyrianpurple.com*

# SHOULD I BUY ....................?

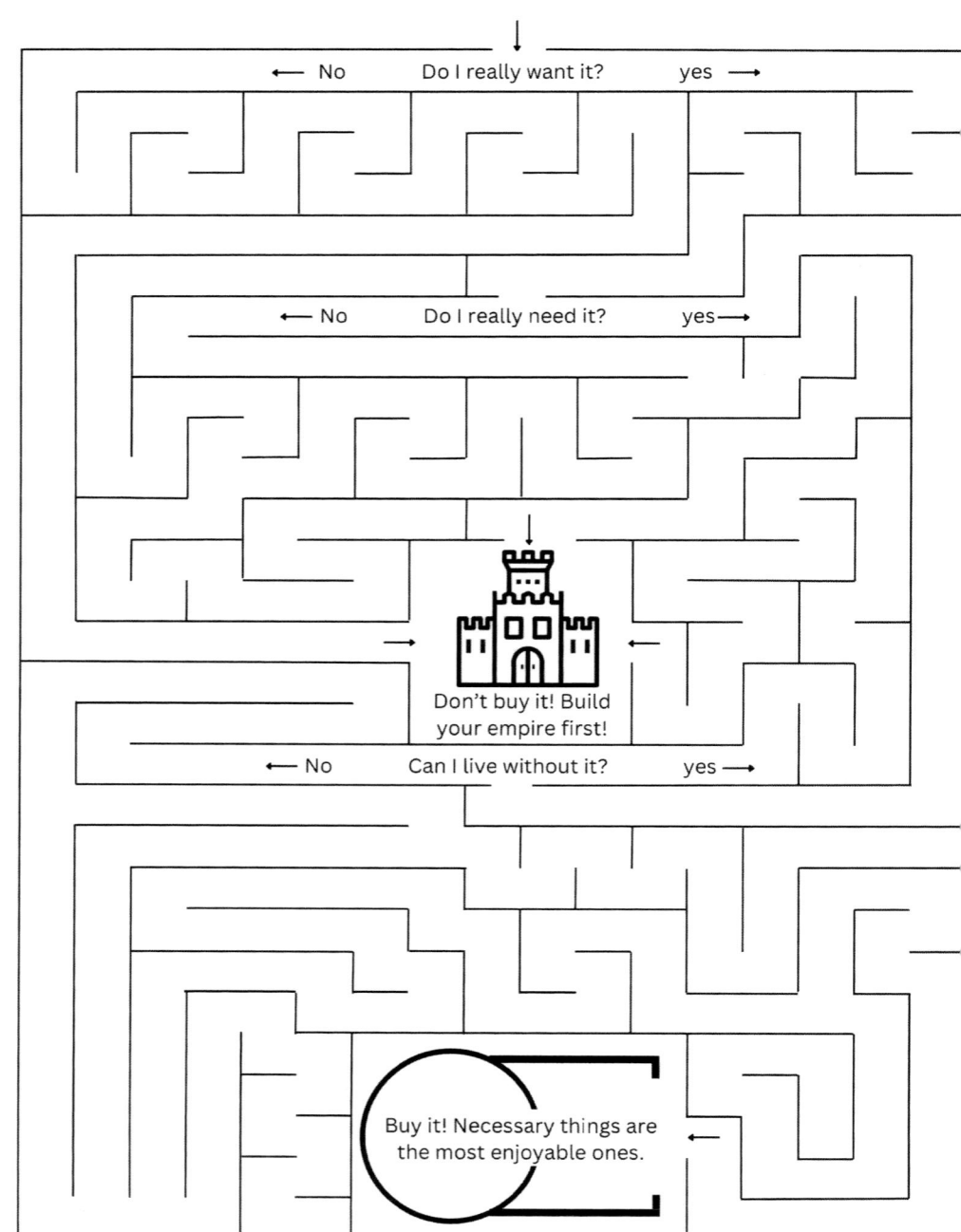

Do I really want it? — No ← / yes →

Do I really need it? — No ← / yes →

Don't buy it! Build your empire first!

Can I live without it? — No ← / yes →

Buy it! Necessary things are the most enjoyable ones.

# Chapter 4 –
# The Richest Man in Carthage

Man sitting on a throne.
Digital painting.

# The Victim Mentality

"You look wealthy!"

Hasdrubal turned back to find a nearly starving dirty child looking up at him. Hasdrubal was distracted by the child while he was discussing business with an assembly of nobles.

He kneeled down and asked, "Are you lost? Do you need someone to show you the way home?"
"No!" answered Rhodian. He was emotionless and continued, "I need someone to show me the way out of misery."
Surprised, but also impressed by his innocence and directness, Hasdrubal said with a little hesitation, "I probably... know someone!"

By now, the father had caught up his son and immediately apologized. "Please forgive my son. He doesn't mean to be rude." Hasdrubal smiled and said, "No worries. You both look like you have a story to tell. Be my guest and join me in my house. It is on the hill just over there."

Gisgo and Rhodian accepted the invitation. As soon as they arrived, a crew of servant took care of the two guests. Firstly, they took a bath, and then they got new robes. Father and son were bid to take a seat at the table on the patio until the generous host joined them. A soon as the guests reached the patio, they were overwhelmed by the wonderful view of the Mediterranean

Sea*. The sun was shining and the bright blue sky was cloudless. A fresh sea breeze was blowing infrequently.

All the fascination disappeared as soon as the food was served! The father and the son couldn't hold themselves back. Hot seafood soups, stuffed fish and a whole grilled leg of lamb was on the menu. They started to devour the food like they had never eaten before. As soon as they sat back, Hasdrubal appeared.

"I guess you liked the food," joked Hasdrubal.
"Yes, sir!" was the short reply from the shy guests.
"Well, could you tell me your story?" asked the amusing host.
The father told the whole story. Hasdrubal listened attentively and asked for details that the father hadn't even told his son. Finally, the father finished the story by saying, "Now we are looking for someone to tell us the secret to getting rich."

"I guess I am the man you are looking for," was Hasdrubal's reaction. He added, "There is no secret behind it. Everyone knows how it works."
"We obviously don't, as you heard," responded the boy, enthusiastically. He then added, "Please tell us how."
"There is an old, famous Babylonian clay containing seven gold rules. However, knowing the laws is not sufficient. Using them is the tricky part. Few people can apply them. I talk based on my personal experience."

"Weren't you born rich?" asked both guests curiously, at the same time.

* Visit Sidi Bou Saïd https://maps.app.goo.gl/pbQo7jH2SzpMhm2FA

"No! I built my fortune all alone," answered Hasdrubal with pride. He continued laughing, "I was poor long ago, as you are now! Well, let me tell my story, too. I was a young boy with two brothers. One was younger and named Himilco and the other was older and called Machaeus.

One sunny day we met accidently in the city. There was a great celebration for the newly elected Sofetes. We knew that there will would be a lot of rich people attending the ceremony. We were hungry and hoped to get some food or something more or less valuable.
Our greed didn't pay off. Disappointed, we sat down in the shadow of Carthage's walls. My young brother Himilco started to complain, saying the well-known idiom:

*New horses but the chariot is the same.*

He thought that those people were spending a whole fortune, but not giving us anything. Machaeus also had the same point of view and blamed them for our misery. He believed that they were taking the whole wealth of the city. Both brothers started to discuss the possessions of the rich."

"I heard that they can afford watermelons during winter seasons!" affirmed the father, Gisgo.
"That's funny!" smiled Hasdrubal and continued, "My brother Machaeus was always saying that. I can tell you something even stranger, some of them keep a black swan in their houses."
"A black swan? I can't even imagine it." The little Rhodian was astonished.

"This is the victim mentality," said Hasdrubal and explained, "Poverty feeds on this mindset. Poor people are always blaming others for their poverty although it is their fault."

"How is that?" inquired Rhodian.

"By turning their strengths into weaknesses," answered Hasdrubal. He then continued, "Me and my brothers were turning our abilities into disabilities. However, we knew someone who had succeeded in doing the opposite."

"Who is that?" asked Gisgo.

"It is Balthazar, our childhood friend. During the ceremony, we discovered that he became a lifeguard despite his deafness."

"What a lucky man!" commented the father. He added, "Probably, he knew someone in order to get that position."

"My brothers shared the same opinion as yours," said Hasdrubal. He continued, "They also thought he had been engaged out of pity. My brothers believed in the saying,

*play it crazy to survive.*

In reality, fighting without hearing noise is a great advantage for a warrior. Balthazar knew that and turned his weakness into a strength. He focused on learning how to fight. While Balthazar turned his disability into a gift, me and my brothers turned our abilities into disabilities."

"How did you and your brothers become rich then?" wondered the little Rhodian.

"It is only me who did that, actually. I will tell you how and why, but first promise me to get rid of the victim mentality!"

"Promised," replied father and son, both at the same time.


**Some of the hidden lessons:**

*~The victim mentality~*

- The victim mentality feeds poverty.

- A strength can be turned into a weakness.

- Turn your weakness into strength.

- Putting the blame on others prevents you from progress.

*~Your own lessons~*

- ……………………………………………..………………………..

- ……………………………………………..………………………..

- ……………………………………………..………………………..

# Ten Winters Later

It was already sunset while Hasdrubal sat with his two guests on the patio. He continued telling his story, "At the celebration day, I told my brothers that we could not keep on going the way we had been for all those years. I informed them about a Babylonian clay that supposedly held the secret to wealth[2f]. It consisted only of seven rules of gold.

Himilco, my young brother, felt that these rules might make sense, but Machaeus was unconvinced."

"Why was he unconvinced?" asked the father, Gisgo.

"He mocked it. He didn't understand why only a few people get rich although the rules are very simple. He considered poverty to be a destiny and that we didn't deserve to be rich. Of course, I didn't agree. I wanted to try the rules. I believed that I deserved to be rich. Machaeus assumed that I am arrogant. He pretended that I considered myself better than him. He attacked me and we fought again. That was the moment where we followed different paths."

"That's sad! But did you apply the rules of gold? Is that how you made your wealth?" asked Gisgo again.

Hasdrubal stood up and said, "I took the Babylonian rules of gold to heart. I started to practice them blindly. Every time I doubted, I said to myself: I don't know a better way than this." Hasdrubal walked and moved his hand indicating to his house, "Bit by bit, I built more than what you see here. More than the half the ships which you can see in the port are mine. I also own land that you can't see from here, but it is as long as a full day ride.

I have even developed the Babylonian gold rules to fit the needs of our days. Now, I will list the Carthaginian gold rules to you, so listen carefully:

1. *Get rid of the victim mentality*
   Verify your mentality. Stop putting the blame on others and take responsibility. Get ready to go through any circumstances, no matter how hard they may be.

2. *Control your expenses*
   Control is knowing how much you earn and spend. It is having limits for your expenses, but also tracing and evaluating expenses and earnings.

3. *Search for and start a business as soon as possible*
   Searching for business opportunities takes practice. The sooner you start, the better it is.

4. *Get the required skills*
   The ability to negotiate, to say no and to understand numbers are essential skills for any business. You should acquire them.

5. *Seek the knowledge of professionals*
   Even if you are familiar with a business, you have to seek the knowledge from professionals.

6. *Make it self-ruled*

Improve your business so that it requires less and less of your effort.

7. *Learn and repeat*
   Learn and repeat the whole process no matter whether you fail or succeed.
Now you know the way out of misery."

Gisgo and Rhodian were very grateful and said, "We will never forget these rules. We will follow them blindly as you did before. But, may you also please tell us what happened to your brothers?"

"I met them ten summers after the fight, when a strange man visited me," Hasdrubal explained, "He wanted to borrow money for a horse farming business. I called the stable boy, who took care of my horses, to survey the discussion.
He spotted that the man was a fraud and asked him some questions that made him uncomfortable. I also discovered that this fraud man had already borrowed money from Himilco, my young brother. I immediately made my way to Himilco in order to warn him. When I arrived, I found my older brother Machaeus there too."

"How did the long-lasting reunion go?" asked Gisgo.
"It was a long meeting," answered Hasdrubal, "By the appearance of my older brother, it was clear that he was still broke. Both couldn't oversee how much my situation had improved. I warned Himilco. I said that I had caught the man with the horse farming business and then I left. As I was leaving, both my brothers held me back. Himilco wanted to thank me.

My advice saved him twelve months of his hard-earned savings. I also wanted to know how they were both doing. I discovered that Himilco started to apply the Babylonian rules two years ago. Machaeus was there trying to borrow money from Himilco. He also asked me if I found a treasure. Obviously, he was still obsessed by this nonsense."

"Did you tell them how you became rich?" wondered Gisgo and Rhodian.

"Yes! I told them everything in detail." Hasdrubal sat down again and continued narrating, "I still remember the look on their faces when I told them that I am *the richest man in Carthage.* They were very curious to know the secret. I will tell you as I did with them. I started by getting any labor I could find. I worked in fields under the burning sun and in stables despite the stench of the animals. At the next full moon, I received my well-deserved earning. Following the rules of gold, I should save a tenth of my earning. And keep doing that for at least the next twelve months. Only then could I start making it multiply."

"For this reason, you kept working very hard," commented Gisgo.

"Yes, I did. But not just that." Hasdrubal smirked and carried on, "If I wanted to make gold multiply, why should I wait so long? I said to myself. I believed in the saying,

*take your part from the start.*

The sooner I was able to multiply my savings, the more money I could get out of it. I had only one problem."

"You didn't know where should you invest your money?" said Rhodian, eagerly interested.

"Well spotted!" Hasdrubal smiled and continued, "The tenth of my earning was only a piece of silver. That was not enough to invest it in trading ships or caravans. I thought it had merely the value of a chicken. Chicken! That word reminded me of an idea that I had when I was a child. I thought once of buying a chicken and selling its eggs.

So, I decided to start my own chicken farm. I bought my first chicken the next morning. However, the following day, I discovered that it was a stolen chicken. Its real owner came by and took his animal away. I continued working twenty-eight days to be able to buy the next chicken. On the second try, what I bought was not a chicken. When I said that, my brother Machaeus laughed at me and asked if it was a goose."

"A goose is even more valuable than a chicken," commented Gisgo.

"That's right!" confirmed Hasdrubal and kept talking, "I ignored Machaeus's ironic question and kept retelling like I do now. What I bought was a young rooster. It was the moment when I learned,

*man makes mistakes until he'll be wise.*

I kept the rooster anyway. It was very frustrating to get tricked more than once, but the rooster was pleasant company. It woke me up early. That made me one of the first available reapers for fields owners. It also enabled me to go early to the market, before the best merchandise was sold. That's how I bought a good chicken.

At the fourth full moon after I started saving, I was relieved to see that I finally had more than a tenth of my earnings and the rooster. I had a chicken with twenty-eight eggs in addition. I decided to buy another chicken, eat eight eggs and brood the rest. The next full moon I was overwhelmed. I got sixteen chicks and twenty-eight eggs besides the two chickens, the rooster, and my earnings. I kept doing that until I was unable to count neither the eggs nor the chicks[1b]. It didn't take so long until I started to sell eggs, chickens and roosters. As they say,

*little money with little money turns to big money."*

"Did you make all your fortune through only farming chickens?" asked Gisgo.

"Surely not!" negated Hasdrubal and went on, "As I often worked in fields, I became familiar with it. I bought and sowed the best spots. That's how I harvested plenty of wheat. One winter, when Baal Hammon seemed to be furious, the sea remained rough. The ships didn't arrive or leave the port. The caravans in winter were less operated anyway. The city lacked commodities. My chicken and wheat fields turned to be literally worth gold. So, I sold them all. To keep my fortune safe from thieves, I bought the most precious thing in the city."

"Did you buy the port of Carthage?" asked Rhodian.

"Nice guess!" laughed Hasdrubal and continued, "I was not that rich either, but that was much better than Machaeus's guess. He asked if I bought jewels, but that would not protect my fortune from thieves. I bought fishermen's ships and cruisers instead. They were at their lowest value due to the long-lasting bad weather. Now, while I am sitting with you, my wealth is still growing like a field of wheat. When I told all this to my brothers they appreciated what I had achieved and bid for help to do the same."

"Did you helped them?" asked the guests
"Himilco just needed good advice. I taught him some skills and sent him to Hanno, the great navigator. Machaeus, on the other hand, had a long road to go."
"I guess it is worth it," said Gisgo.
"Unfortunately, his story didn't have a happy ending," commented Hasdrubal sadly.


**Some of the hidden lessons:**

*~Hasdrubal and his brothers~*

- For some people, poverty is a destiny. For others, it is a choice.

- When you try to change people close to you, they will either follow you or be hostile towards you.

- You don't need to convince people to change and sometimes it is better if they don't know that.

- Some people will never take you seriously until you succeed.

*~The richest man in Carthage~*

- Use the Carthaginian rules of gold to build your wealth.

- Think big start small.

- It is never too late.

- Consider failure as part of the success and keep trying.

*~Your own lessons~*

- ………………………………………………..……………………………..

- ………………………………………………..……………………………..

- ………………………………………………..……………………………..


**Tools for Practice**

Use the following template to track your expenses: Find out which expenses are an investment, a necessary expense, or just an instant pleasure. Try to find a new source of income. Think about how you can turn active income into passive income...

*You can order this template and many other items at*

*www.thetyrianpurple.com*

# EXPENSES CONTROL

Goal: <u>Save at least the tenth of my earning</u>　　Month: <u>January</u>

## Income

| Description | Time | Amount | Active/ Passive |
|---|---|---|---|
| Salary for working from 9am to 5pm | 40 hour/week | 2000$ | active |
| Side line | 4 hour/week | 400$ | active |
| Car sharing | 1 hour/week | 200$ | passive |
|  |  |  |  |

## Variable/fixed expenses and bills

| Date | Description | Amount | Investment | Instant pleasure |
|---|---|---|---|---|
| 01.01 | Drinks and snacks | 17$ | No | Yes |
| 01.01 | Groceries | 56$ | No | No |
|  |  |  |  |  |
|  |  |  |  |  |
|  |  |  |  |  |
|  |  |  |  |  |
|  |  |  |  |  |
|  |  |  |  |  |
|  |  |  |  |  |
| Total |  |  |  |  |

## Recap

| | Goal | Actual | Difference |
|---|---|---|---|
| Earnt |  |  |  |
| Spent |  |  |  |
| Debt |  |  |  |
| Saved |  |  |  |

Evaluation:

# EXPENSES CONTROL

Budget Goal: _____     Month: _____

## Income

| Description | Time | Amount | Active/Passive |
|---|---|---|---|
|  |  |  |  |
|  |  |  |  |
|  |  |  |  |
|  |  |  |  |

## Variable/fixed expenses and bills

| Date | Description | Amount | Investment | Instant pleasure |
|---|---|---|---|---|
|  |  |  |  |  |
|  |  |  |  |  |
|  |  |  |  |  |
|  |  |  |  |  |
|  |  |  |  |  |
|  |  |  |  |  |
|  |  |  |  |  |
|  |  |  |  |  |
|  |  |  |  |  |
|  |  |  |  |  |
| Total |  |  |  |  |

## Recap

|  | Goal | Actual | Difference |
|---|---|---|---|
| Earnt |  |  |  |
| Spent |  |  |  |
| Debt |  |  |  |
| Saved |  |  |  |

Evaluation: _____
_____
_____
_____
_____
_____
_____
_____
_____
_____

# The Poor Pharaoh

Under the Mediterranean stars, both guests, Gisgo and his son Rhodian, gazed in fascination at Hasdrubal.

"Why didn't you just give your brothers a bunch of gold?" wondered the father.

"I was convinced from the beginning that man must

*sweat to appreciate the bread.*

Let me tell you the end of the story and you will understand the reason," answered Hasdrubal. He carried on, "We agreed that we would meet the following day at the port. The next morning, Machaeus arrived very late. But this was not a big problem because the ship we were waiting for arrived even later. I introduced Machaeus to the crew, who were happy to have more manpower on board.

My brother's first task was carrying the crates of fish from the ship down to the port. I clearly saw discomfort in his face. He liked neither the smell of the port and fish nor the task itself. The thing that he disliked the most was being told what to do. After work, the crew arranged together to go back to the sea the next day, before dawn. Later, I was informed that my brother didn't join them."

"Did something bad happen to him?" wondered Rhodian.

"No, he came to me two days later." Hasdrubal continued, "He pretended he had seasickness. I was upset because I believe that man can *dig a well even with a needle.*

All he needs is *faith, endeavor* and *patience*! I had some understanding that my brother might really suffer from seasickness. Therefore, I engaged him to work with some diligent reapers in the wheat field. The very next day, he came back to me."

"He seems to be lazy, what brought him this time to you?" asked Gisgo.

"Yes, I also noticed that," confirmed Hasdrubal and continued to talk, "He complained this time about the burning sun. I told him that he must first deserve what he earns. Out of envy, he asked why this didn't apply to me. Why money was raining on me at home, while he was burning outside under the sun. I explained why I didn't need to drop a single bead of sweat. The reason is that I sweated enough already. I sweated until I made it look ridiculously easy. I simply made money multiply itself. I reminded him of the Babylonian rules of gold."

"Did he show understanding?" asked Gisgo, while the little Rhodian fell asleep.

"Yes, he was convinced," affirmed Hasdrubal. He carried on, "To reward him for his understanding, I made him supervisor of a group of reapers. He worked continuously for two full moons. Three days after payday he asked me for three coins of gold. I wanted to know what happened to his last payment. He told me that he used the money on women and wine. I insisted on knowing about the previous payments, but I always got the same answer. He was poor yet living like a pharaoh. He was *beggar and swagger.*

Since the gold was allegedly for paying his last debt, I gave it to him. After all, he was eventually responsible for his behavior. I tried to keep him working for me by paying him in advance. My intention was to teach him how to manage his money. That was my disastrous mistake."

"Why was that the case?" asked Gisgo.

"Soon, I found out that Machaeus had another plan. He used the money to pay three bandits to rob a caravan. These robbers killed a man of a highly respected family. Unfortunately, one of the bandits was caught. He led them to the rest of the crew and all of them were sentenced to death." Hasdrubal finished his story.

"That's sad," commented Gisgo as Rhodian woke up.

With a burning heart, Hasdrubal sighed, "They say give a man a clay of wisdom and purse full of gold. He will then ignore the wisdom and lose the gold. My brother didn't just ignore the wisdom and lost the money, but he also lost his life. I misjudged Machaeus and learned a very bitter lesson:

> *Man makes mistakes even if he is wise*
> *man makes mistakes until he dies."*

"We are sorry for your loss," said Gisgo and pleaded, "We are not

> *looking for a cold bread.*

We need your assistance, like your brother."

All of a sudden, a servant interrupted them and said, "Master Hasdrubal! The general Hannibal and his assistant are here to meet you."

"Alright, I will meet them in the royal suite," answered Hasdrubal and said to Gisgo, "I have a spices store and I need you both to oversee it. Tomorrow, we will talk about all the details."

Hasdrubal finished the discussion and took care of his new important visitors.

**Some of the hidden lessons:**

*~Poor pharaoh~*

- Bad money management is the most common reason for poverty.

- Some people never change.

- When there is no will there is no way.

- Money can help solve problems, but it can also cause bigger problems.

- Embrace mistakes as new lessons.

- Some shortcuts ensure that you never reach your destiny.

*~Your own lessons~*

- ………………………………………..…………………………..

- ………………………………………..…………………………..

- ………………………………………..…………………………..

# Chapter 5 - Carthaginian Ceramic

Carthaginian ceramic.
Graphite drawing.

# The Soul Healer

It was midnight and the full moon lit up the calm Mediterranean Sea. The wave-less bay reflected the wonderful moonlight. It nearly turned the night into day. In the atmosphere, silence prevailed, except for the yelling owl.

Rhodian was rolling over in his bed for the thousandth time, trying to sleep. The owl's hoot was so loud, as if it was inside his room. He simply couldn't sleep. "What is the reason for being awake at this late time?" he started to wonder, "Is it the owl? Or is it the full moon?" After a short while thinking, he said to himself, "For the last two weeks, even without owl and moon, I have felt the same as I do now!" He added, "It seems to be something else."

The night was finally over and Rhodian was far from recovered. Nevertheless, he was obliged to go very early to the market. He followed Hasdrubal's saying,

*no vendor, no labor.*

Rhodian supervised their spices business with his father Gisgo. They bought it from Hasdrubal after spending years working for and learning from him.

Rhodian arrived at the same time as his father. There were a few clients waiting already. Father and son opened the store and started immediately to work. Whilst serving the clients, Rhodian was half asleep. He barely could keep his eyes open. When they finished the orders, his father noticed that his son was very tired so he asked him to search for a healer.

Soon after, a caravan with new merchandise from a faraway desert arrived. Gisgo called the leader of the caravan. "Did you bring what I ordered?" The caravaneer went to him and said, handing him a rare spice, "How could I forget! Here are your grains of paradise."

"Hey! You are safely back! Tell us about the journey!" said Rhodian, awake.
"Oh young man, you look horrible! What happened to you?" laughed the leader of the caravan.
"I can't sleep at night," Rhodian answered in a low voice.
"I told him to search for the healer," commented Gisgo.
"I'm afraid that the healer can't help," said the leader of the caravan, "It's true that he knows how to heal the body. However, he can't help with sleep problems. Such an illness is soul related." He kept silent for a moment then continued,

*the owner of the soul is its healer.*"

A momentary silence enveloped all three men, until Rhodian broke the silence - "I don't think I can be a healer in any manner."

The caravan leader smiled and said, "I maybe don't know how to heal your soul, but I think I know some tools that you can use. Bring me a cord, two cloths of different colors and five small stones."

Rhodian started directly to collect the stones. His father looked for the cloths and the cord. The caravaneer ate almonds in the meantime.

Gisgo and Rhodian handed the man five small stones, a short cord, one black cloth and one white cloth. The caravan leader cut the cord into two pieces and said, "Look Rhodian! The first thing you have to do is to think deeply and find out what you really love." He tied the cloths with the cords and made them into two little bags. He then handed it all to Rhodian and continued to explain his instructions.

"The second thing, ask yourself if you are fulfilling your dream or at least fulfilling your duty! Ask yourself five times a day: when you wake up, at sunrise, at noon, at sunset and before you go to sleep. Whenever your answer is 'yes' put one of the five stones in the black bag. If you are not fulfilling any of them, then put one stone in the white bag." Rhodian looked at him with wide-open eyes. He waited for more instructions, but they never came. Rhodian then asked, "Is that all? What will happen if I do that?" The caravan leader walked away saying, "I think that you've understood the instructions. I must take care of my caravan now!"

After work, Rhodian sat at his favorite place, near his home, under an olive tree over the hill. Holding the bags and stones in his hands, he spent the whole evening thinking. He tried to find answers to the caravan leader's questions. He discovered that the only thing he really loved was pottery. Rhodian loved watching his favorite pottery master while he was working. He could now apply the caravan leader's instructions.

The following morning, Rhodian woke up late again after sunrise. He quickly prepared to go to the market. Before leaving, he remembered the stones and the little bags. He asked himself, "Am I fulfilling my dream or at least my duty? I guess none of them! Following the caravan leader's instructions, I have to put a stone in the white bag because the answer is negative. Not only one but two, one for the first prescribed time and also for the second." He then put two stones in the white bag and left.

While he was working, his father asked him, "Did you find out your dream?" Rhodian was a little bit shy to say the truth. Pottery had nothing to do with selling spices. His father might then ask a lot of questions which would make him feel uncomfortable. Instead, Rhodian answered, "I am not sure yet, but thank you for reminding me." After work, just before midday he visited his pottery master. When he arrived, the pottery studio was closed. "What a pity, but at least I tried! I can finally put a stone in the black bag." thought Rhodian.

Rhodian made his way to the yard and watched some similar-aged guys wrestling. The fights were very exciting, nevertheless, Rhodian was tired so he fell asleep. At sunset, one of the last spectators woke him up. He decided to go home. While he was collecting his belongings, he saw the bags and asked himself, "Am I fulfilling my desire or my duty?" without answering he put a fourth stone in the white purse and made his way home.

Late at night, he went to bed. He didn't seem to be tired. Jumping from thought to another, he remembered the caravaneer and asked himself, "Am I fulfilling my desire? No." He stood up and put a stone in the white bag.

At that moment, the white bag contained four stones while the black one contained a single stone. Holding both bags in his hands, he started to wonder, "What will happen now? And how can I heal my soul?" Rhodian had neither answers nor sleep. He decided to try something different. The next day, he would make sure that all five stones would end up in the black bag.

At the first rooster crow, Rhodian stood up immediately. He wanted to visit the pottery master. He put a first stone in the black bag and hurried to the pottery shop. When he arrived, the pottery master was not there yet. It was too early.

At sunrise, the master arrived. He found the excited young man waiting in front of the shop.
"Good morning master!" Rhodian greeted him.
"Good morning! Nice to see you Rhodian. What brought you here so early in the morning," wondered the master.
"I would like to spend the whole day working with you," answered the young man.
"Alright, I have a lot of work for you. Many heavy tasks are waiting for young power," the old man smiled happily.

Rhodian put a second stone in the black bag and started to work. He transported heavy bags of different soils, broke and sifted clay. He enjoyed watching the master create wonderful jars and pots. Furthermore, he also tried vainly to build something.

76

At midday, the master closed the shop and left to go to a place where he could find special minerals for his profession. Rhodian put a third stone in his black bag and accompanied him the whole afternoon. After a day filled with new experiences and lessons, both men bid goodbye. On his way home, Rhodian took some clay raw material with him. He arrived home at sunset, so he put the fourth stone in the black bag. Even at home, Rhodian worked enthusiastically with the clay until he felt tired. After putting the last stone into the black bag, he threw himself in bed and fell deeply asleep.

The night passed by, the rooster started crowing, but Rhodian didn't notice it. The sun shined, but he kept sleeping. After midday, a loud knocking on the door woke him up. His father was looking for his son. When Rhodian opened the door, Gisgo hugged him and said, "Thank goodness you are fine! Yesterday I thought you overslept, but today I was worried about you." Rhodian answered with half-open eyes, "No, I'm fine! I actually feel great!"

The father was surprised and asked, "Did the caravan leader's advice work out?"

"I'm not sure, but all I know is that I slept well." answered Rhodian.

"Alright, I'm happy to hear that! We have a lot of work today." said his father.

"Okay, I'll join you later." Gisgo left and Rhodian sat and wondered about the previous day. It appeared that if he didn't pursue his dream, his spirit would be in pain.

The decision was clear. Rhodian wanted to let down his family business and work for the pottery master, even for free.

Some of the hidden lessons:

*~People are vital for business~*

- Finding your life path needs deep thinking.

- Keep track of your actions to avoid losing your life path.

- The right people are very important for keeping the business running.

- Your dreams and duties may heal or sicken your soul.

- You don't need to tell everyone about your intentions.

*~Your own lessons~*

- ………………………………………..…………………………..

- ………………………………………..…………………………..

- ………………………………………..…………………………..

**Tools for Practice**

The following template helps you to track your day:

Identify and write down your tasks and dreams. Check every hour during the day to see if you are meeting any of these. Draw an ascending arrow if the answer is yes, if the answer is "No," the arrow should be drawn descending

*You can order this template and many other items at www.thetyrianpurple.com*

# THE SOUL HEALER

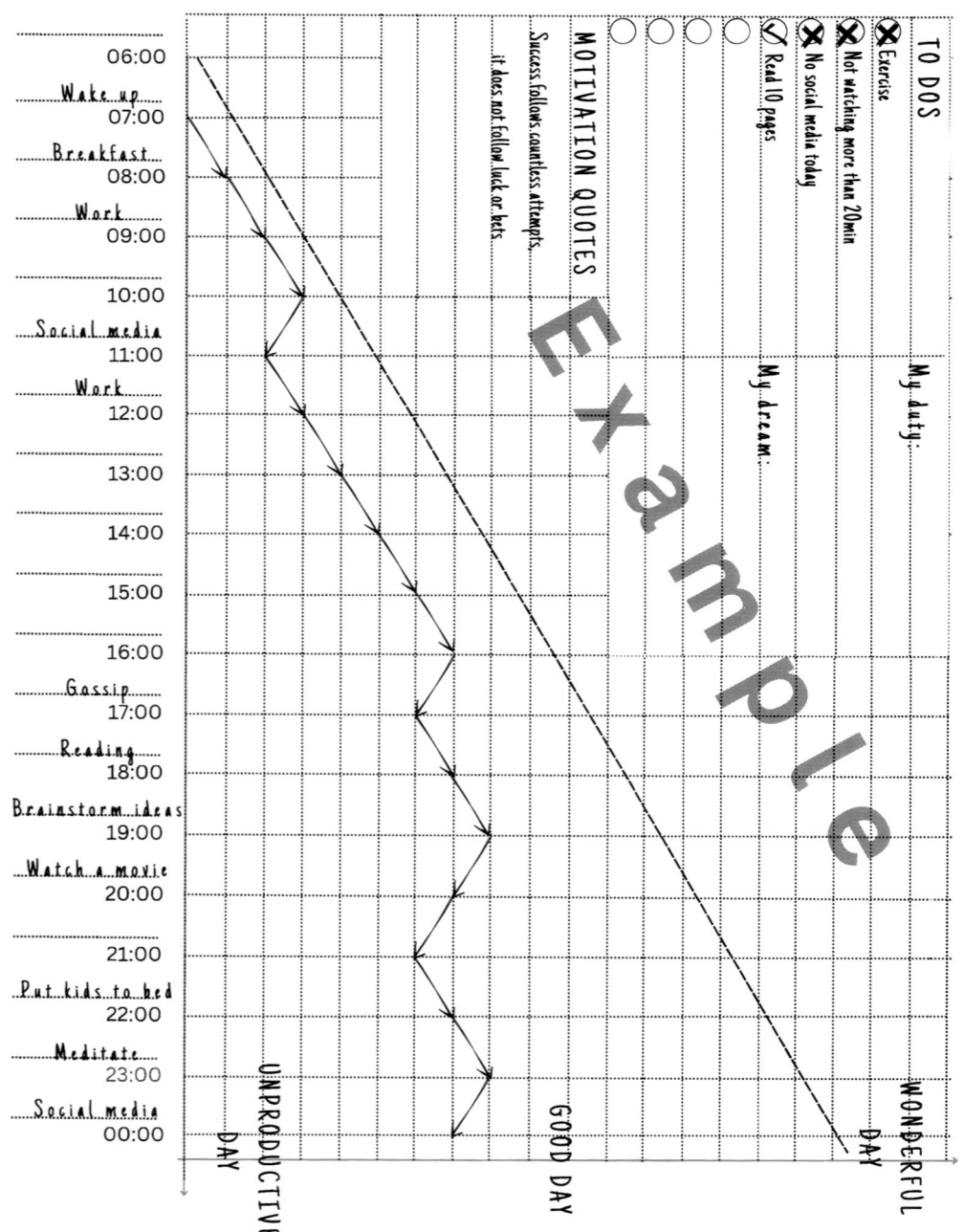

**TO DOS**

- ⊗ Exercise
- ⊗ Not watching more than 20min
- ⊗ No social media today
- ✓ Read 10 pages
- ◯
- ◯
- ◯
- ◯

My duty:

My dream:

**MOTIVATION QUOTES**

Success follows countless attempts,
it does not follow luck or bets.

Time schedule (left axis):

- 06:00
- Wake up — 07:00
- Breakfast — 08:00
- Work — 09:00
- 10:00
- Social media — 11:00
- Work — 12:00
- 13:00
- 14:00
- 15:00
- 16:00
- Gossip — 17:00
- Reading — 18:00
- Brainstorm ideas — 19:00
- Watch a movie — 20:00
- 21:00
- Put kids to bed — 22:00
- Meditate — 23:00
- Social media — 00:00

UNPRODUCTIVE DAY

GOOD DAY

WONDERFUL DAY

# THE SOUL HEALER

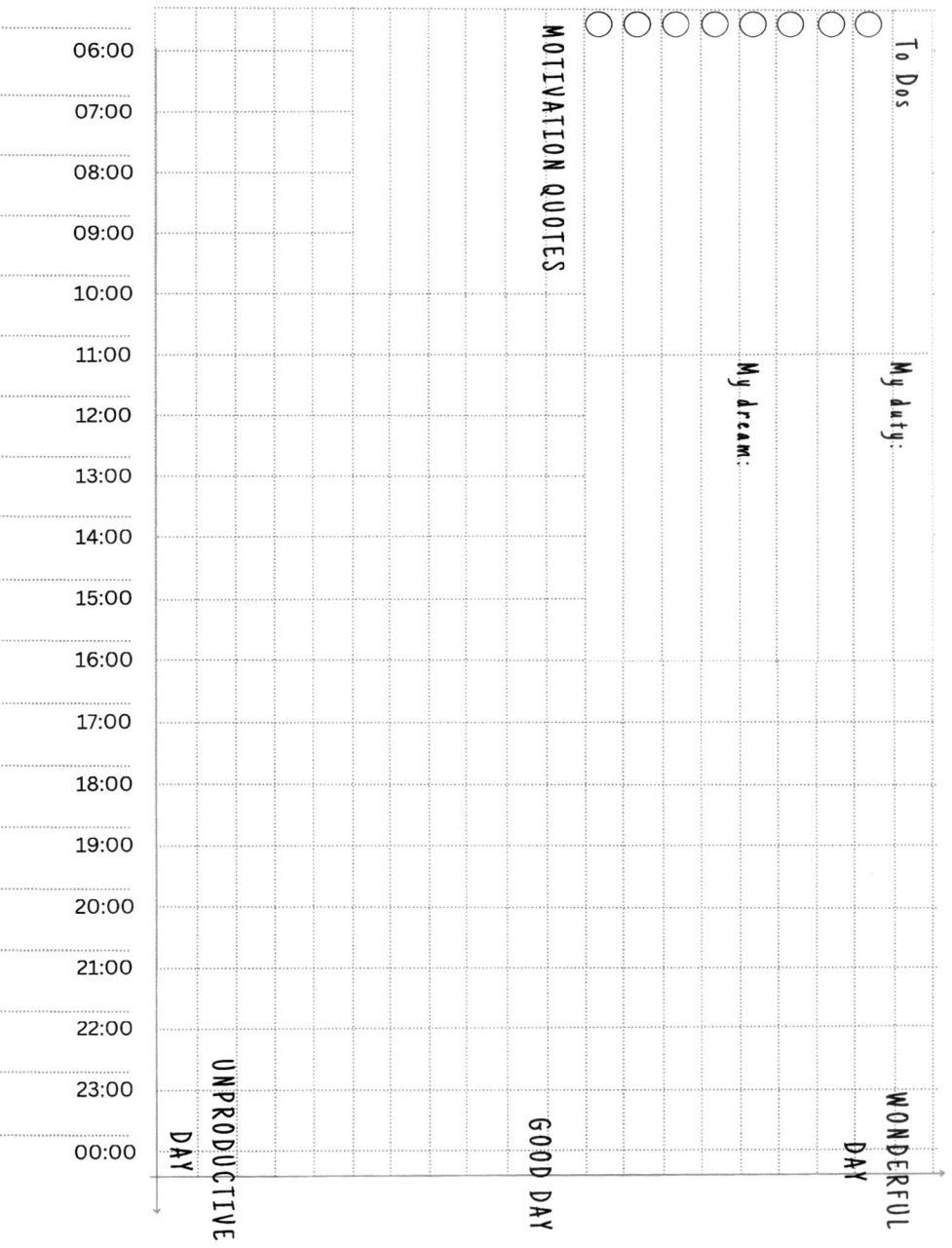

06:00
07:00
08:00
09:00
10:00
11:00
12:00
13:00
14:00
15:00
16:00
17:00
18:00
19:00
20:00
21:00
22:00
23:00
00:00

MOTIVATION QUOTES

To Dos

My dream:

My duty:

UNPRODUCTIVE DAY

GOOD DAY

WONDERFUL DAY

# The Existential Change

It had been a long time since Rhodian started working for his pottery master. Rhodian took over the pottery shop, after his master passed away. Throughout the years, he even became better than the master. People bought his unique products like fools. They waited for months to get what they ordered. He even exported products abroad. Rhodian's family became one of the wealthiest families in Carthage.

The neighboring shops knew that his business was flourishing. They tried to copy his products. Sometimes they were even successful. Over time, the whole town* became known for pottery. It was the main place where any kind of pottery could be found (simple pots, pots that fit inside each other, pots with covers, etc).

Although Rhodian didn't need to work anymore, he kept doing it as a passion. However, the years passed by and he was no longer making any innovations.
There were rumors among the people that the painted potteries were easier to sell than the rest. Rhodian didn't like the idea at all. It didn't make any sense to him. Why should he paint his potteries?

Rhodian tried to convince his neighbors to continue using traditional methods. Nevertheless, over time, the neighboring shops began selling almost only painted potteries. Rhodian

82

insisted on continuing with the old methods. This was the reason that his client numbers significantly decreased.

One day, Rhodian sat in his shop and called his student, "Hey Bomilcar, come over here, we have something to talk about." The diligent and loyal worker went to him and his master continued, "You were always helpful and more importantly you always were and still are trustworthy."

"It is my job, master Rhodian," said the humble student.

"I was wrong, trying to change salesmen instead of changing myself.

However it is too late now," confessed the master.

"It's not the end of the world. We still can change," replied Bomilcar.

"I don't even need to do this anymore, but your living depends on it. That's why I've decided to offer you this location. It is a compensation for your hard work throughout these years," said Rhodian.

Bomilcar didn't believe what his master had just said. Despite his joy, he knew that the current business had limited value. The other pottery shops had already won most of the clients. For this reason, the student had an idea that had the potential to change the whole business. He bent over his master and whispered in his ear.

When he finished, Rhodian looked impressed and said, "You got my shop and still want my money?"

"For a good purpose! People still need a last lesson. It is for Rhodian's reputation!" said Bomilcar.

"Alright, on a condition that you give me one-fourth of the earnings. You may know that I already made enough money out of it. My intention is to reassure future income for my descendants," said the master.

"Alright, but you have to pay my expenses as long as I don't make any earnings. I will then give you the one-fourth of the earnings every full moon," proposed the student.

Bomilcar and Rhodian shook hands and made a deal.

For three months, Bomilcar came and left without making any sales. Every day, he brought minerals and water to the shop. He mixed them up in different ratios to create unique paint colors. Repeatedly, he tested the paint on the pottery.

Often, he was sarcastically asked by the neighboring salesman about the failing business. Sometimes the student answered, but most of the time he focused on painting and ignored them. He kept working and made multiple attempts. Every time he was satisfied with the result, he saved the invented paint colors and exposed the colored product.

One day, while he was working in deep concentration, a noble woman passed by. She was very attracted by his displayed colored product and addressed him impressively, "I have seen a lot of colored pottery but these colors! They are so vivid! I have never seen these combinations before. It is so exceptional." The woman fell in love with his product. She bought it for a purse full of gold and left.

Bomilcar was filled with joy and kept working hard. Thereafter, two high-class women visited his shop and asked for the same product. He presented them instead with different products. The women were fascinated by the products' beautiful colors. They had forgotten about what they had originally came for and bought them immediately. Word spread in the neighborhood. The other salesmen envied him and wanted to know the secret of his success. They tried to copy his colors in vain. It was only Bomilcar who had the color recipe.

Few days later, the most successful pottery master came by and asked Bomilcar to supply him with precious pottery paint colors. It didn't require much time for the remaining prosperous pottery masters to start buying the colors from Bomilcar. The business was back again.

Informed of the success of his student, Rhodian decided to pay him a visit. As he arrived, he found his student looking messy with colors. Bomilcar greeted him and said, "Master, the plan was fruitful. I changed our business. Now, we simply sell unique pottery paint colors instead of pottery products." Rhodian proudly smiled at him, and commented,

*do like your neighbor or change your labor."*

## Some of the hidden lessons:

*~ Change is existential~*

- Work not only for money. Work for experience and passion.

- The one who can't change or can't change fast enough has bad cards.

- Ensure more than your own income. Ensure a future income for the people valuable to you.

- Humans are survival artists, they not only want to stay alive but they also want to survive financially.

- The key of competition is to possess what others don't have.

*~Your own lessons~*

- ……………………………………………..…………………………..

- ……………………………………………..…………………………..

- ……………………………………………..…………………………..

# The Challenging Soul

Rhodian sat in the painting shop and started a conversation with Bomilcar, "Despite your success, I can still feel that you are tired and upset. Tell me, what do you have in mind?"

"I must confess, I can't sleep at night, although I'm doing what I love," said Bomilcar.

"No need to worry," smiled Rhodian and continued, "Your soul will always search for new challenges. I can only suggest an exercise for you. Before going to bed, list the thoughts that are disturbing you. Name a pot after each topic and make sure they have different colors. Collect as many small stones as you can. Then just relax and try to free your mind."

"I can't. I always think about something," disagreed Bomilcar.

"That's right and it's essential for the exercise," confirmed Rhodian, then added, "Every time you notice yourself starting to think, identify the topic and put a stone in the related pot. Then stop thinking and say to yourself - Not now! Tomorrow, I'll take care of that - and relax. The more you practice, the sooner you stop thinking. Keep doing that until you feel sleepy, then go to bed."

"What if I wake up at night and start thinking?"

"If you can't sleep, then get out of bed and start the whole exercise again. The following day, order the pots by the number of stones they contain. Find out which pot has the most stones and take action towards the related topic. Once you have completed your task, you can celebrate by smashing the pot on the floor."

"That's a strange method! From where do you know it?" wondered Bomilcar.

Rhodian smiled and left saying,

*"He who was born one night before, possesses one trick more."*

## Some of the hidden lessons:

*~Bomilcar~*

- Your wanting will always exceed your having.

- Expressing yourself relieves.

- You are not the first one who has your problem. It is highly recommended that you ask the right person for help.

*~Your own lessons~*

- ………………………………………..…………………………..
- ………………………………………..…………………………..
- ………………………………………..…………………………..

**Tools for Practice**

The following template helps you to deal with overthinking: Identify the thoughts that are disturbing you and group them into topics. Name each jar after a topic. You can paint the jars with different colors.

Try to free your mind. Every time you notice yourself starting to think, identify the topic and cross a stone in the related jar. Then stop thinking and say to yourself - Not now! Tomorrow, I'll take care of that - and try to free your mind again.

once you finish, write on the back of each jar the todos required to deal with the corresponding topic. The todos of the jars with the most crosses could be done first.

*You can order this template and many other items at*
*www.thetyrianpurple.com*

Payments/Debts

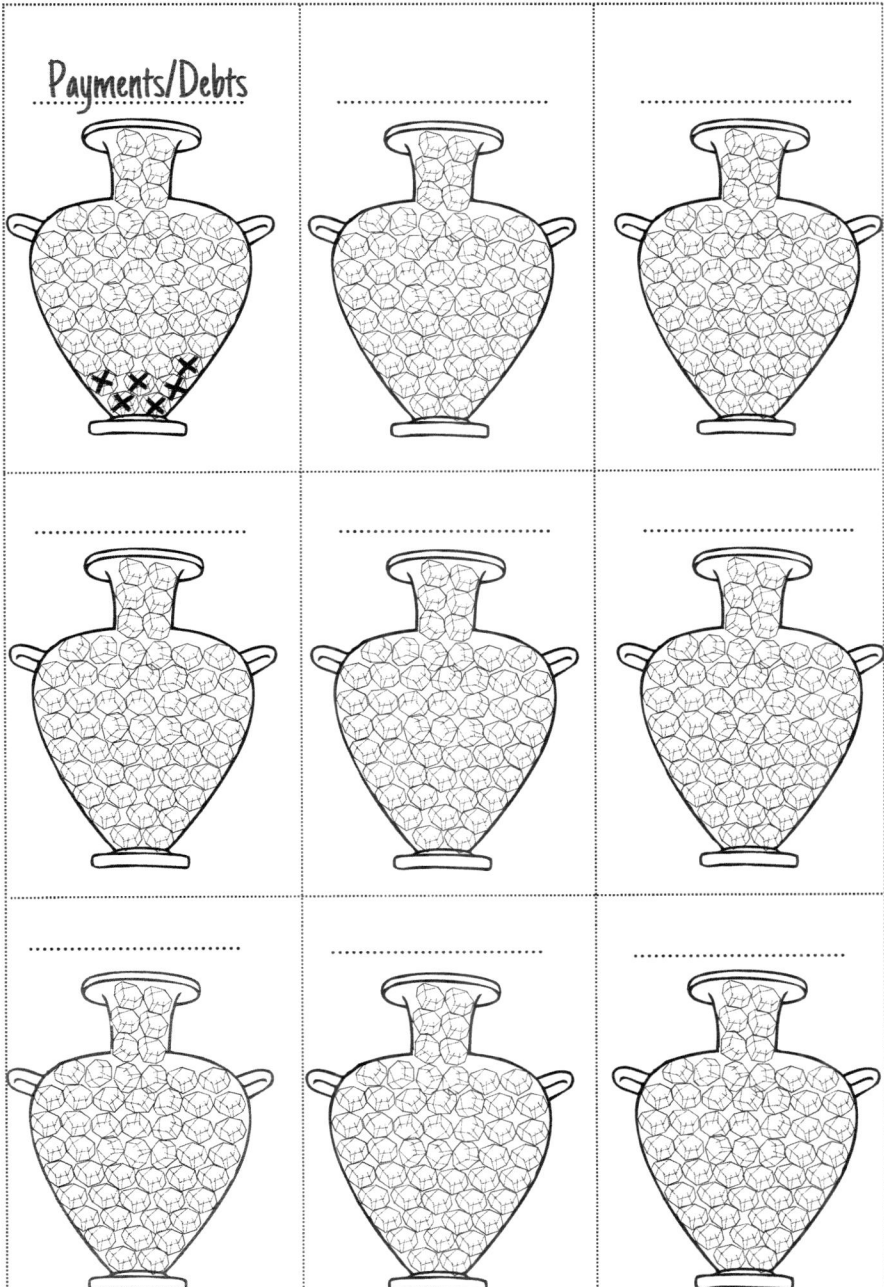

Consult a financial advisor

Read financial articles/books

Find new sources of income

Develop business ideas

# Chapter 6 – Hanno and Himilco

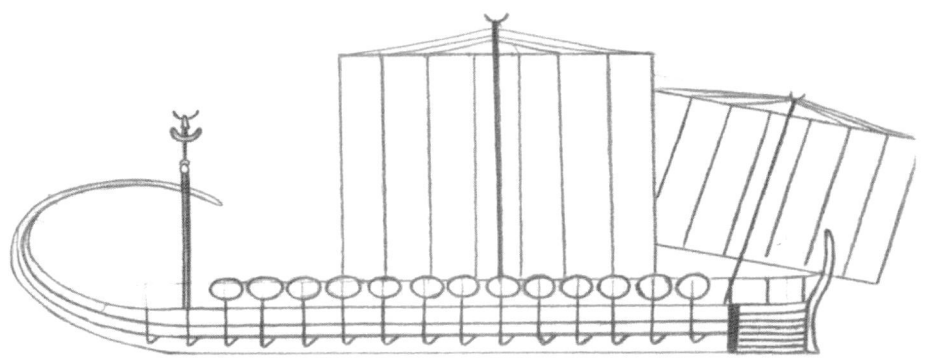

Carthaginian ship.
Graphite drawing.

# Glass for Diamonds

"You can count on me!"

shout Hanno to the pedestrians. He was the great navigator and best sea trader of Carthage. He was trying to get funds for the next sea trading journey. Usually he didn't need to do this. Fundraisers used to approach him without his call. Unfortunately, in those days fundraising was becoming riskier.

The reason was that Roman ships were dominating the eastern side of the Mediterranean Sea. People looked away and kept managing their own affairs. Disappointed, Hanno went back to his ship. He was not really convinced by the undertaking. He laid for a while thinking about what to do, then fetched something to eat.

While he was eating, a starving cat slowly approached him. Hanno started playing with the cautious cat. Holding a piece of dried fish, he invited her to eat from his hand. He called her, "Come on, here is my offering! Come on,

*offer and forget suffering or suffer and forget offering.*"

Hanno continued to say it over and over again as Himilco got on the ship.
Himilco learned the essential skills for business from his brother Hasdrubal. Although most people didn't like Hanno, Hasdrubal recommended him as a trustworthy business partner. In fact, people were just misled by his bad manner.

94

Hanno noticed the presence of the unexpected man. At first sight, he knew that Himilco was a reaper. This man couldn't fundraise the amount that Hanno needed for his sea trading journey. That's why he ignored him and kept playing with the cat.

"Captain Hanno! Hasdrubal has guided me to you," Himilco interrupted his game and continued, "I heard your speech today! I believe in you and I want to put my money on you. Here are one hundred pieces of gold!"
Hanno couldn't ignore him anymore. He turned and said, "Not bad for a reaper! What profit do you expect from this investment? This doesn't even cover the supplies that we will need for the journey."
"Yes, I know, and I still insist on having one third of the earnings," said Himilco adamantly.
"No way! This man is absolutely crazy,

*by the first blare he wants his share.*

The whole trading costs about 1000 coins. You can get one tenth of the earnings at most!" denied the captain.
"I can understand that you are angry. In addition to the 100 coins, I will help you convince the remaining fundraisers," insisted Himilco.
"How would you do that? You are probably just

*selling wind to the ships."*

Hanno underrated Himilco.

"I'll let you know how, but tell me first, do we have an agreement?" asked Himilco.

"No! I can't suppose that you will be funding one third. That would mean more than 300 coins. By giving me your 100 coins I can consider you to be funding 150 coins," suggested Hanno.

"I wish I could accept that. What about one forth, let's say 250 coins, and you head in any direction you want. Deal?" Himilco was improvising.

"Wait,

*wool is sold for its heaviness.*

I would like to head to the west instead of the east. If you could convince the fundraisers to do that, I would consider you to be funding 175 coins," suggested the merchant.

"Now we are getting closer, let me think about it," answered Himilco. He started to think, then after a moment said, "Consider me in funding 203 coins, I can't accept less than that. As a sign of my good will, I can add a bag full of wheat as an additional fund."

"Why three coins? What difference would a bag of wheat possibly have?" said Hanno, who hung on Himilco's words.

"That's my last offer! Should we give up on this? Suffer or offer… remember?" finalized Himilco.

"Now you are speaking my language!" smirked the captain.

"Alright, here is my plan." Himilco started to reveal how he would convince the fundraisers. Both men agreed to meet the next day at midday in the downtown. It was the time when the city was the most crowded.

"Glass for gems!" screamed Himilco. People started to stop and listen. Himilco continued, "Glass for gems is what each one of you desires. This exchange is waiting in faraway lands at the west side of the sea!"

"Who are you to talk about merchandise? You are just a reaper!" asked someone from the crowd.

"Yes, I am a reaper! It is not me who is going to navigate. It is the most known captain in Carthage! The only man who has been there and returned safely. It is Captain Hanno!" answered Himilco and pointed at Hanno who was also standing in the crowd.

"The west has unpredictable weather! This is a risky offer," assured another man.

"Yes, it is! Every fundraiser knows that there is always a risk. But keeping your money also has risks. It can be stolen. One thing is also sure, it will get taxed by the government. Those who cling to their money will watch bitterly as their wealth slowly gets squeezed by the government. He who wants his money to multiply would know that this a once in a life time opportunity."

People began touching their purses, but they needed one last incentive.

"May everyone witness that I will grant my whole earnings of two winters to Hanno!" yelled Himilco, throwing a purse of gold to Hanno in front of everyone. "He has limited the number of the fundraisers to ten. This offer is available just for three days. Just nine places left. If you want to make a deal with Hanno, you know where you can find him." Immediately, one member of the crowd gave one hundred gold coins.

"May everyone know that from now there are only eight places left," shouted Himilco as he took the money from the man.

Himilco and Hanno went to the ship and the interested fundraisers started to come by. The next morning, a wealthy man paid the rest of what was needed for the new adventure.

The undertaking took six full moons before Hanno could return. People envied Hanno, Himilco and the rest of the fundraisers for their unprecedented gains. They completely forgot how Hanno had pleaded with them and how they had not accepted to risk their money.


**Some of the hidden lessons:**

~*The negotiation between Hanno and Himilco*~

- Negotiation is a main skill for any business.

- "No" may mean "I am not convinced" or "not now" but never means "no".

~*Himilco to the crowd*~

- Influencing people is an essential skill, especially for fundraising.

*~The crowd and the cat~*

- There is no business without risk.

- If you want to make your money multiply you have to take calculated risks.

- Never look back.

*~Your own lessons~*

- ………………………………………..……………………

- ………………………………………..……………………

- ………………………………………..……………………


**Tools for Practice**

This template can assist you in either staying committed to your choice or reconsidering it. It can also help you avoid doubts and regrets.

Go through the following steps every time you think about making a life-changing decision:

1. Cut the card into 6 pieces.
2. Fill the pieces as follows:

In the "Intention" section, write down what you want to achieve with your decision.

In the "Purpose" section, write down the reasons that lead you to your decision.

In the "Consequences" section, write down the most important possible consequences.

In the remaining parts, mention at least one risk or disadvantage of your decision.

3. Build the puzzle.

4. Take each puzzle piece and read what's behind it.

5. Put it back if you agree or set it aside if you disagree.

A complete puzzle means you are ready and should make this life-changing decision.

An incomplete puzzle means you are not yet ready for this life-changing decision. Maybe you should reconsider it or look for other alternatives.

You can save the puzzle for future use in case you feel doubts or regrets about your decision and you want to remember the factors that influenced your decision.

*You can order this template and many other items at www.thetyrianpurple.com*

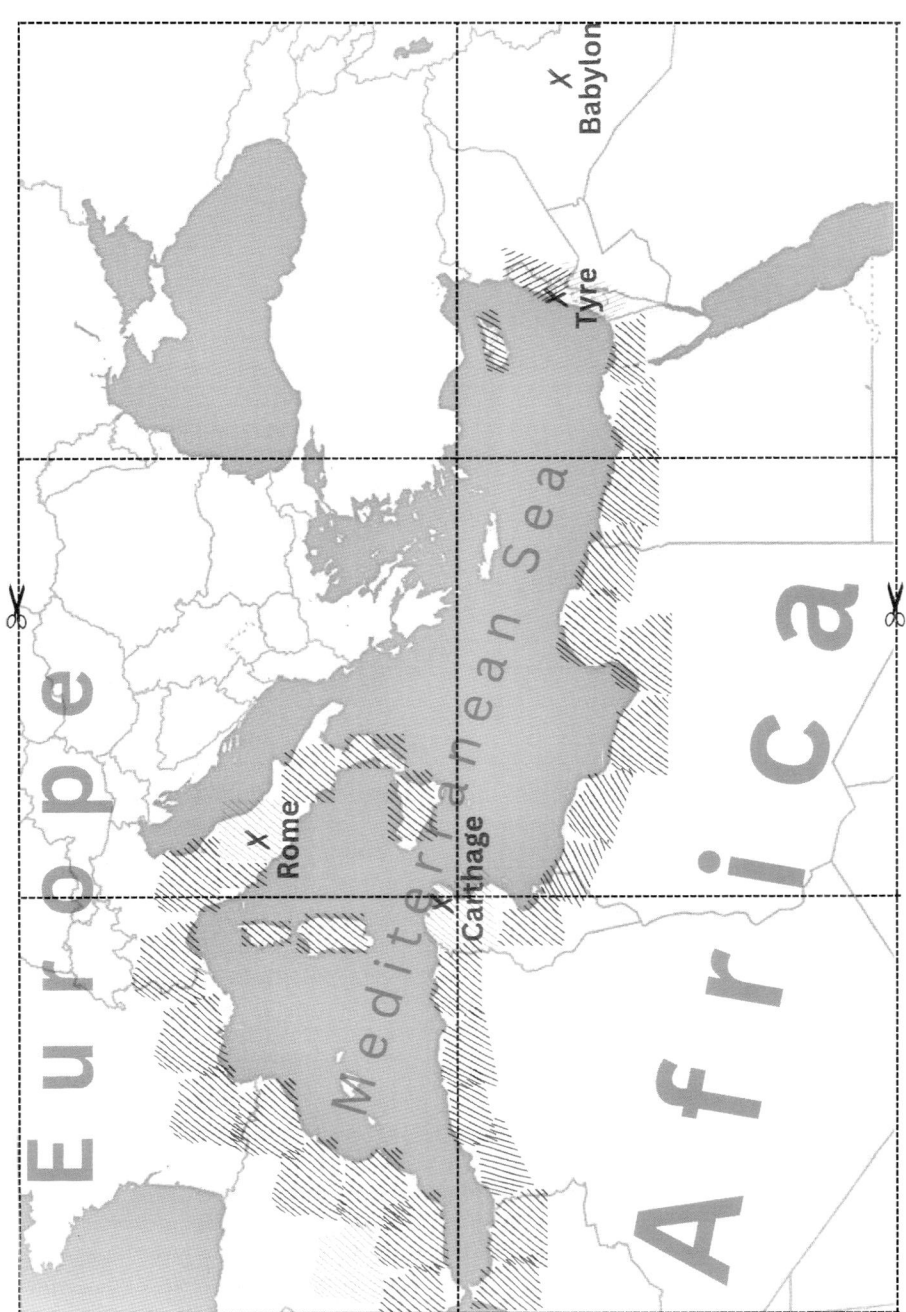

## Intention

What I want to achieve is:

_____
_____
_____
_____
_____
_____

## Purpose

I made my decision for the following reasons:

_____
_____
_____
_____
_____
_____

## Consequences

I am prepared to face the following consequences:

_____
_____
_____
_____
_____
_____

## Risk/Disadvantage

I stick with this decision even if:

_____
_____
_____
_____
_____
_____

## Risk/Disadvantage

I stick with this decision even if:

_____
_____
_____
_____
_____
_____

## Risk/Disadvantage

I stick with this decision even if:

_____
_____
_____
_____
_____
_____

## Intention

### What I want to achieve is:

my Financial Freedom

## Purpose

### I made my decision for the following reasons:

Helping weak and poor people

Spend more time doing what I love

To have and enjoy more time with the people I love

## Consequence

### I am prepared to face the following consequences:

give up the current job

## Risk/Disadvantage

### I stick with this decision even if:

... if I have to work without getting paid

## Risk/Disadvantage

### I stick with this decision even if:

I might have to work all week

## Risk/Disadvantage

### I stick with this decision even if:

Even if I will often be forced to go
on business trips

# The Last Feather

"Hanno, the captain is back!"

Different people rushed around and informed each other. The port filled with people who were standing and watching the horizon. What they were staring at slowly became clearer. It seemed to be a small ship heading for the port. The closer it approached to port, the bigger it became.

The great merchant ship had finally reached the port. People watched what the famous Hanno brought with him this time. The seamen started to unload spices from far lands, elephants' tusks and wild animals in big cages. He even brought treasures, acquired from pirates who had tried to rob his ship.

Most people were expecting Hanno to appear wearing heavy gold necklaces. Unexpectedly, he came out with his standard look. However, on his shoulder stood a marvelous parrot. The exceptional beauty of the bird fascinated the people. Its colorful feathers were so intense. The people wondered: "What a wonderful bird!", "Why doesn't it fly away?", "From where did he get it?"

Hanno directly made his way to his favorite tavern. In there, his fundraisers started to come by. They were the people most interested in his arrival. He owed them a lot of money. Somehow, he had to pay all his seamen, the loyal men who had accompanied him for months, through stormy weather. They had multiple skills. Some of them knew different languages and most of them knew how to fight against robbery.

Himilco was the first who paid him a visit. Hanno gave him his share right away. Himilco wanted to talk about the journey but the wonderful bird at Hanno's shoulder stole his attention.
"How did you catch it?" Himilco asked.
"I didn't! We crossed many oceans until we reached a warm land. The bird's parents left the nest, as we cast anchor at a magnificent beach. I took care of it, gave it shelter and fed it. It considers me as its parent now," answered Hanno.
"What's your price!? I want to buy it from you! Just tell me how much you want," demanded Himilco, so thrilled.
"It is priceless! This bird has a magical ability. It can talk!" The captain denied Himilco's request and smirked.
"Like us humans?" asked Himilco, doubtfully, "I can't believe it! I want to see and hear that!"
Hanno signed and the parrot clapped his wings yelling, "Hanno! Hanno! Hanno!"

Himilco was speechless and insisted on buying it. Hanno disagreed and gave him one of the parrot's feathers instead. The parrot didn't like that, but Hanno was its owner. While Himilco was leaving, another fundraiser came by. He saw the beautiful feather in Himilco's hand. He asked for a feather too. Hanno didn't want to loose his fundraisers. That's why he accepted and answered, "Yes, you can have one."

The fundraisers kept coming and going. They were jealous of each other. All of them wanted to own at least a parrot's feather. Indeed, every time someone asked, Hanno responded with, "Yes, you can have one." After a few times, the parrot learned the answer and began to reply instead of its owner.

This kept going until the parrot had only one long pretty feather left. Hanno wanted to please an important fundraiser, who hadn't come yet. When the next person asked him for the last feather, the parrot automatically said, "Yes, you can have one," but Hanno refused and said, "No, I am sorry!"

The parrot was shocked to learn that it could say, 'No'. He discovered new dimensions by having a choice to refuse. After a few rejections, the parrot learned quickly to say 'No'. The parrot's lesson was that

*'No' prevents troubles.*

Two days later, the important fundraiser came by and asked for the last feather. "No, I am sorry!" replied the parrot. Nevertheless, Hanno took it by force. The parrot was so angry and hurt that it decided to leave his owner. Featherless, it tried to fly away. It fell immediately to the ground. Hanno took his bird from the ground and put it back on his shoulder.

The parrot was rethinking his relationship with Hanno. It thought, he

*neither loves me nor can he resist me.*

It was confused, especially by the fact that it was possible to do things in different ways.

The following day, Hasdrubal, the richest man in Carthage, came by and asked for one feather. Hanno promised to give him one feather as soon as they grew back, in twenty-eight days.

Right before the last day, the parrot flew away. Its journey started, but the troubles also began. People began chasing it. Predators were attacking it. Hungry and thirsty, it hid in a tree. Just before it gave up, it saw a herd of parrots landing on a neighboring tree. The parrot joined the group and finally lived the life of parrots. The risks were still there, but the parrots were always singing,

*"Hanging with the group is a holiday."*

The parrot finally had no need to please someone who took his feathers one after the other, in a pejorative way.

**Some of the hidden lessons:**

*~The parrot's keeper~*

- During your evolution, someone who initially supported you could end up becoming your greatest obstacle.

- Saying no is a learnable skill even for an animal.

*~Your own lessons~*

- ………………………………………..…………………………..

- ………………………………………..…………………………..

- ………………………………………..…………………………..

# Chapter 7 - Carthaginian Marble

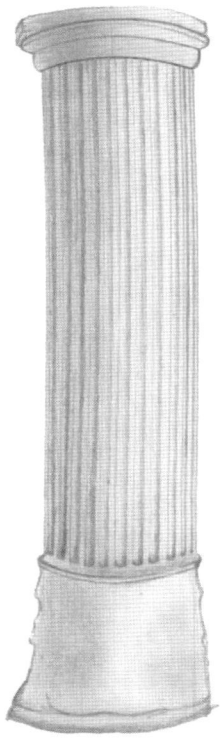

Carthaginian pillar.
Graphite drawing.

# Chisel The Legend

"Say that you are afraid of me!" threatened Bostar in a provocative manner. He was hitting the same orphan child again. The pitiful child was weaker and younger than him. For an unknown reason, the child didn't want to say that. The other children started to laugh at Bostar. His ego was hurt, but he couldn't do anything. He expelled the child from the yard. With a black eye and a bloody nose, the child made his way home. He was not able to prevent himself from crying.

"Why did you hurt him?" asked Mago, one of the children.
"Don't pity him! His uncle is a stonemason. Do you know what he earns for each ordered figure? More than what my dad would earn, if he would work the whole twelve moons!" answered Bostar.

*"Real men earn, and millers yearn."*

said Mago in a low voice.
"Just shut up or you will join him!" shouted Bostar at him.

The child went to his uncle. After telling him what had happened, the uncle took some leather and started to cut it. Soon, he finished and handed it to his nephew. "Is this a weapon? Can I defend myself with it?" asked the excited little child.
"Yes and no. It is a slingshot. You can shoot stones with it. I have made it for you, so you can play without needing the other kids," answered his uncle.

"Thank you uncle!" said the child, and left.

The next day, the child met Mago in a different yard. Mago felt guilty and told him, "I am sorry for what happened yesterday. Why didn't you just say what he wanted?"

"If I were to do it, I would make him happy even though he hurt me," explained the child spontaneously adding, "Anyway, it doesn't matter now."

"What do you have in your hand?" asked Mago, curiously.

"It is my new toy. I can hit things with it from long distances. I am training. Would you like to train with me? If we master this, we can use it to hunt birds!" replied the child.

"Great! Let's do that," Mago agreed.

Both kids trained while other children started to come into the yard. Some wanted to try and others just wanted to watch. After a while, Bostar also joined them. He was envious again. He

*came from behind and said this is mine.*

First, he pretended that he just wanted to try the toy too, but then he didn't give it back. He lied and said that the toy was his. The child just couldn't let it be. Although he knew he had no chance he started to fight. Again, the little boy began weeping and made his way to his uncle. Losing the toy that he got from his uncle had hurt him more than the injuries.

The child told the story to his uncle and said, "Uncle please! Come with me and punish him!"

"I am like your father, I cannot always be there for you. Think about it. What would you do if I were not there with you? You would defend yourself on your own," comforted the uncle.

"I tried hard, but I can never win," replied the child, desperately. "You were using force, although you are weaker than your opponent. What you need is not force. Let me explain to you. Do you see that marble?" The uncle pointed to a huge block of marble and said, "Take the hammer and try to pierce a hole in it."

"I can't, I don't have your power," said the child, not even trying.

"Even me, I don't do it by brute force. I use different techniques to reach my end aspiration. It needs a combination of experience, tools and techniques. Despite all my efforts, the marble may break sometimes. All I would do then is to try one more time. Out of the broken pieces, I can also build Gods. The secret is to keep trying. Do you know what? Although the marble is so solid, it can be pierced even without any force."

"Without any force? Do you mean by magic?" the child was extra attentive.

"Not by magic but by the countless tries that make achievement look like magic. Do you see the water cask over there," the uncle pointed at a cask in the corner. He added, "It drops water on a piece of marble for years. Take a look at the marble now."

The child took a look and said, "It has holes!"

"That's why we say,

*gurgles melt marbles.*

112

If you don't succeed at something, it means you haven't tried enough. Don't confuse this with trying hard enough. It's not about how hard you try, but how often you try. Do you remember how many times you tried to walk? How many times you tried to talk? You learned to walk and talk because you tried often enough. Not because you tried hard enough. We try hard to shortcut the way to save time or effort. It might sometimes work, but it doesn't always work. The same applies to you. You didn't beat the bully because you haven't tried enough times to beat him. And you didn't use any technique. I told you how I chisel statue and figures. If you apply what I taught you, you will chisel more than that. You will chisel a legend, dear *Hannibal!*"

**Some of the hidden lessons:**

*~The stonemason's advice~*

- Struggles can make you into a  legend or break you into pieces.

- Broken pieces can build more than legends.

- Trying so many times makes achievements look easy, like magic.

- It is not about willpower, it is about continuity.

*~Your own lessons~*

- ………………………………………..…………………………..

- ………………………………………..…………………………..

- ………………………………………..…………………………..


**Tools for Practice**

Now that we know that success is about continuity and not about willpower, talent, or luck, this template will help you track your attempts.

*You can order this template and many other items at www.thetyrianpurple.com*

DATE:

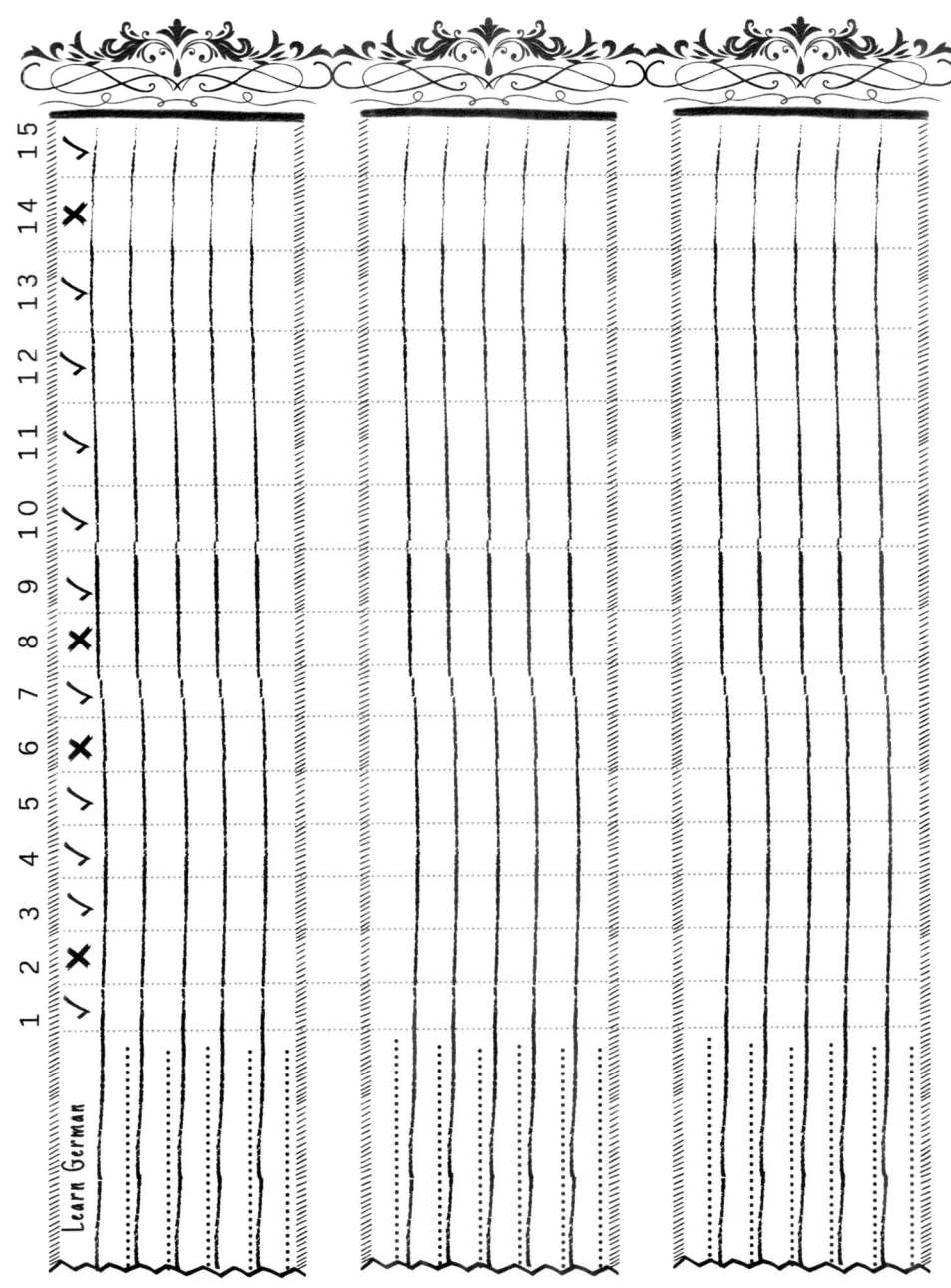

Learn German

1 ✓
2 ✗
3 ✓
4 ✓
5 ✓
6 ✗
7 ✓
8 ✗
9 ✓
10 ✓
11 ✓
12 ✓
13 ✓
14 ✗
15 ✓

DATE:

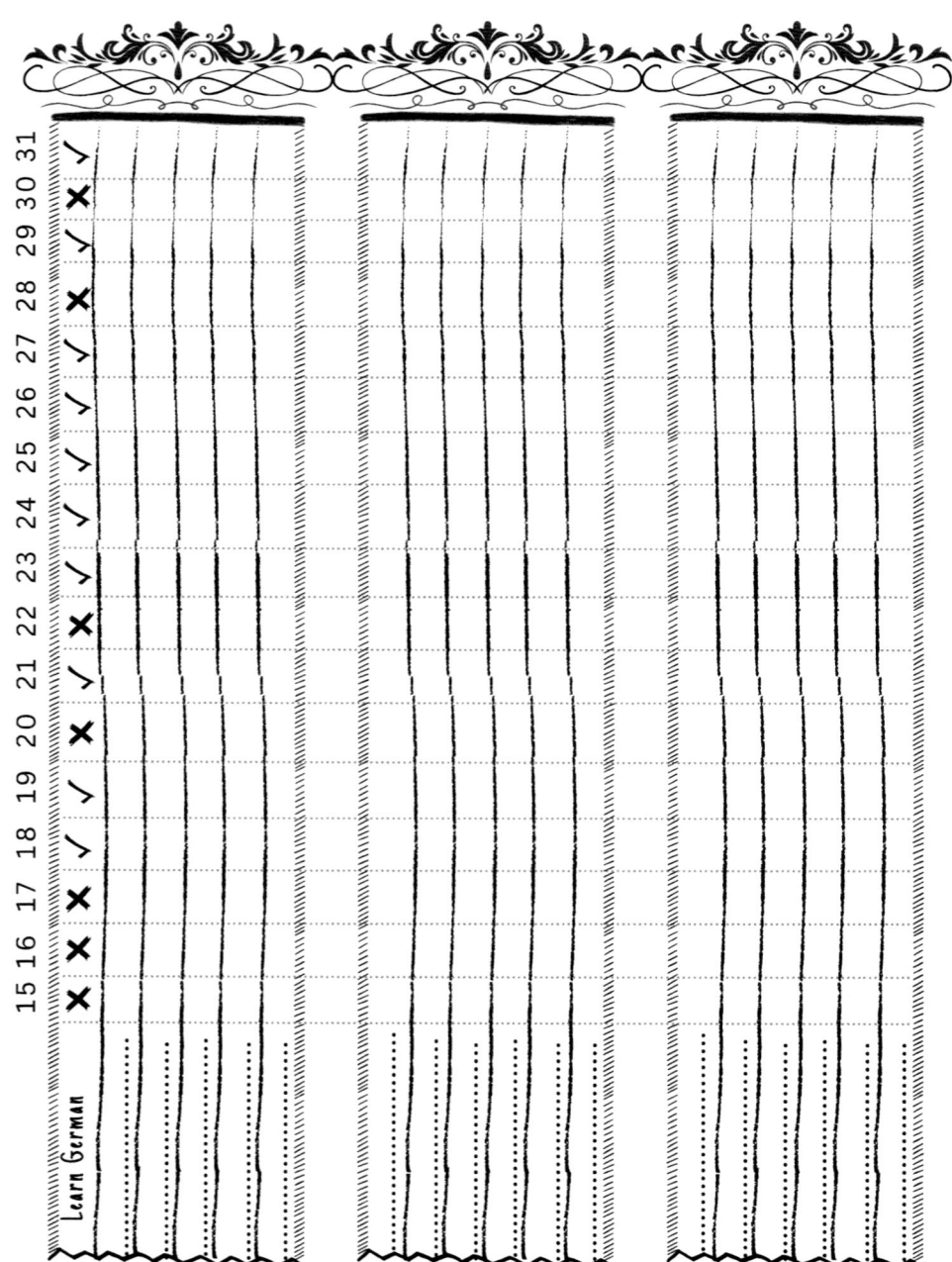

Learn German

15 16 17 18 19 20 21 22 23 24 25 26 27 28 29 30 31

# Chapter 8 - Hannibal

The bust of Hannibal.
Graphite drawing.

# The Striking Skill

Hannibal had to rethink his uncle's lessons for the rest of the day.

The next morning, Hannibal played in an abandoned field. He collected ladybugs in a purse. He saw Bostar battering Mago. Hannibal wanted to defend his friend but remembered his uncle's lesson. Hence, he observed Bostar while he was fighting instead. He noticed that Bostar's strength came from his stable position. Therefore, Hannibal developed a strategy to beat Bostar.

After the fight, Hannibal invited Mago to collect ladybugs with him. Despite his bruises, Mago accepted. It was more enjoyable than being bullied. Immediately, Hannibal told Mago about his strategy. He knew that Bostar would not leave them playing in peace. Mago was hungry for revenge, so he accepted. A short time later, Bostar approached them.

"What do you have in your hands... ladybugs!" Bostar aggressively grabbed the purse from their hands. While opening it, Hannibal interrupted him, "No! Don't! There are lizards inside!" When Bostar was distracted, Hannibal sent a signal to his friend. Mago caught Bostar's feet. Immediately, Hannibal jumped and placed Bostar in a headlock. Bostar couldn't breathe anymore. Confused and not knowing what to do, he began to cry.

After a moment, Mago ran away. Hannibal pushed Bostar away and jumped into a tree. "You will see what I can do! I will tell your uncle!" Bostar was crying and threatening at the same time. Hannibal yelled from above,

*"Beat me to cry! Pass me and snitch!"*

Hannibal stopped him from climbing by kicking him. Bostar started to throw stones. Hannibal tried to hide but was stung by a wasp. There was a wasp's nest at the top of the tree. He picked up a stick. He hit the nest with that stick until it fell on Bostar's face. The wasps attacked Bostar, who fled in pain. Then, Hannibal climbed down and ran away. He was stung by a few wasps, but he had won the fight this time. Despite the pain, he felt joy! For the first time, it was his opponent who was crying, not him.

At home, Hannibal proudly told his uncle everything. He finally found his passion. He loved to fight little baddies. The parents of these children, however, often came by to complain to his uncle. His actions led to injuries in the head, sand in the eyes and even broken bones. The uncle knew that his nephew couldn't follow his path and become a stonemason. He could only be like his deceased father, an army commander. The uncle then proudly said,

*"Son of the mouse becomes a digger."*

He took Hannibal to Phameas. Phameas was an officer who trained young guys to join the army. The officer accepted Hannibal to take part in the training. For Hannibal, fighting in the training arena was very difficult. He was less successful than the other trained fighters. The fights were different from the ones

119

in the street. He didn't enjoy it. One day, he decided not to take part in the training and watch the fights instead. He was able to observe the fighters, understand their techniques, and find out their weaknesses and strengths. As they say,

*the watcher is a warrior.*

The officer couldn't overlook Hannibal's progress. Phameas taught him more and more fighting techniques. After ten years of training, Hannibal was finally able to beat even the older and much more experienced fighters.

One day, Phameas gave his students a new challenge. He pointed to the strongest fighter and Hannibal, made them leaders, and asked them to build their groups. Phameas gave Hannibal's opponent the advantage of choosing his fighters first. The opponent selected the strongest fighters and Hannibal got the rest. Both groups had the same number of fighters, but they were not equal in strength.

Every fighter got a piece of cloth. Each fighter had to get the adversary's cloth without losing his own cloth. The students had to do this while Phameas counted to ten. Some of them bound the cloth around their arms. The others bound it around their feet or even put it inside their underwear. The officer then checked the number of cloths that each group possessed. Hannibal's group didn't have a chance to win. They lost all fights.

For Hannibal, it was one of the worst training sessions ever. He took a walk by the southern seaside of Carthage and kept thinking about the fights. He saw a group of men pushing a small boat. They were trying to move it from the water to the sand. One of them called to Hannibal for help. Hannibal was too kind to reject. He started to push with them, but it was still too heavy.

One of the men stopped pushing, stood in front of the boat, and counted to three. At three, they pushed at the same time. Even with less people pushing, the boat moved as if it were lighter! Thanks to this, Hannibal understood what to do in order to master the difficult challenge. He remembered the first fight with Mago. He learned that,

*teamwork makes heaviness less.*

During the next training session, just before the fight started, Hannibal prepared a strategy with his teammates. They selected the weakest fighter and placed him in the middle of the group. The team members then gave him their cloth, making sure not to catch the attention of the opposing team. The fight started. Hannibal's team built a line while both teams approached each other. The selected fighter chose to hide all the cloths in his underwear and walked unnoticed behind the line (see following figure).

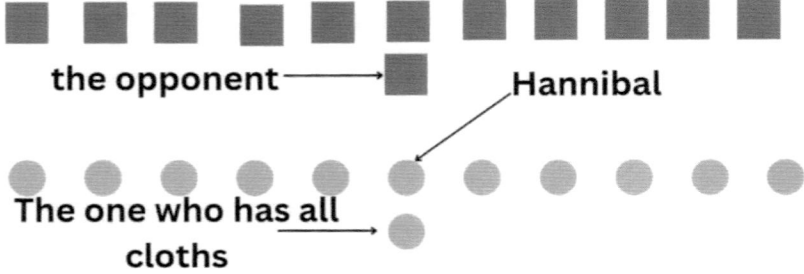

the opponent ⟶ ▪ Hannibal

The one who has all cloths ⟶

When they were close enough, Hannibal's team built a circle around the selected fighter. When the time was up, Hannibal's team had the most cloths. The opponents tried to get the cloths from Hannibal's fighters, although all the cloths were in possession of the selected fighter in the middle.

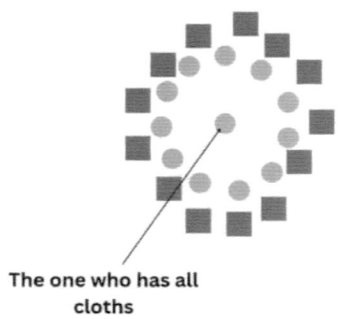

The one who has all
cloths

Phameas was impressed and started a new round. This time, Hannibal's opponent tried in vain to catch the one in the middle. Every time one of the defense fighters fell, the team shrank the circle. Hannibal won again.

A few days later, Phameas was invited by a high ranking commander. The senior commander asked the officer to choose some students to join the army. The commander was planning an attack and needed as many fighters as possible. Phameas invited the commander to show the striking skills of Hannibal. He might be a great strategy assistant, the commander thought. He accepted the offer and accompanied Phameas to the training arena.

The commander was impressed by Hannibal's strategy, but he refused to be assisted by a young man. Phameas insisted. The commander then decided to put Hannibal to the test. He took a few fighters away from Hannibal's team. The teams were no longer equal in number.

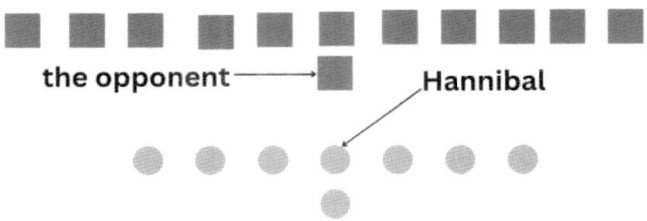

If Hannibal were to keep the same strategy, the opponent would get the selected fighter before the time was up So, he changed his strategy. Again, he secretly distributed the cloths amongst the defending fighters. When the fight started, the opponent focused on the selected fighter who didn't have any piece of cloth this time. Hannibal's team won even with fewer fighters.

123

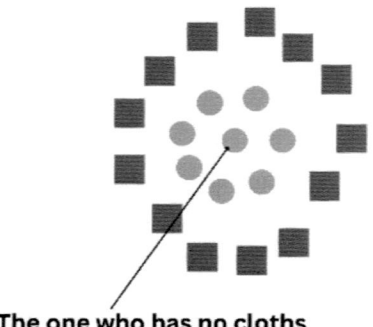

**The one who has no cloths**

The general asked for another round. Maybe he was defending his ego, but he also enjoyed watching the fights. Hannibal had to improvise again. He knew this time that the other team would not blindly attack the selected fighter in the middle. Hannibal picked the fastest two fighters and divided the cloths between them.

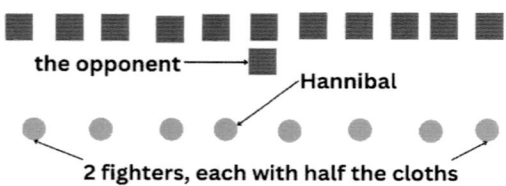

the opponent → Hannibal

**2 fighters, each with half the cloths**

Their mission was to run away every time they faced an opponent. When the fight started, the opponents laughed at them. They thought that they would run away out of fear. They then fought those who were ready to fight. When the time was up, Hannibal's team won the round again.

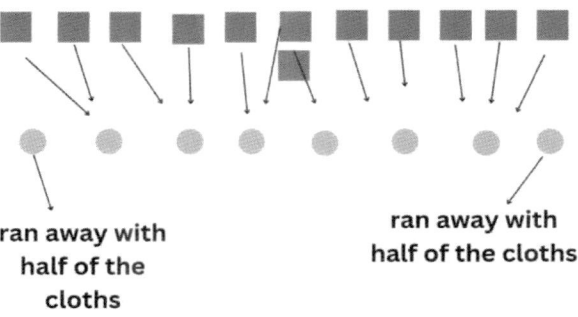

ran away with
half of the
cloths

ran away with
half of the cloths

The senior commander surrendered. He was very impressed and said,

*"small but always amazing."*

I want him. I will set Hannibal as my strategy assistant."


**Some of the hidden lessons:**

*~Hannibal's striking skill~*

- Excellency is not about talent, it is about skills. Skills are learnable.

- A skill will be striking if it matches your ability.

*~Bostar~*

- Ignore those who hurt you and put guilt on you.

- Take a moment to watch and reflect on your efforts, then evaluate and improve.

- Team-work makes the dream work.

- Don't underestimate anything any time.

*~Your own lessons~*

- …………………………………………..…………………………..

- …………………………………………..…………………………..

- …………………………………………..…………………………..


**Tools for Practice**

The following template helps you to reflect on and evaluate your effort.

*You can order this template and many other items at*
*www.thetyrianpurple.com*

# THE WATCHING WARRIOR

Reflection date: Reward:

## What was good? What was bad? What can be improved?

| | Very bad | Bad | Neutral | Good | Very Good |
|---|---|---|---|---|---|
| Focusing on my objectives | ○ | ○ | ○ | ✗ | ○ |
| ............................................... | ○ | ○ | ○ | ○ | ○ |
| ............................................... | ○ | ○ | ○ | ○ | ○ |
| ............................................... | ○ | ○ | ○ | ○ | ○ |
| ............................................... | ○ | ○ | ○ | ○ | ○ |
| ............................................... | ○ | ○ | ○ | ○ | ○ |
| ............................................... | ○ | ○ | ○ | ○ | ○ |
| ............................................... | ○ | ○ | ○ | ○ | ○ |
| ............................................... | ○ | ○ | ○ | ○ | ○ |
| ............................................... | ○ | ○ | ○ | ○ | ○ |

Things that can be done differently next time:

○ _____     ○ _____

○ _____     ○ _____

○ _____     ○ _____

○ _____     ○ _____

# THE WATCHING WARRIOR

Reflection date: _____          Reward: _____

**What was good? What was bad? What can be improved?**

|  | Very bad | Bad | Neutral | Good | Very Good |
|---|:---:|:---:|:---:|:---:|:---:|
| .................................................. | ○ | ○ | ○ | ○ | ○ |
| .................................................. | ○ | ○ | ○ | ○ | ○ |
| .................................................. | ○ | ○ | ○ | ○ | ○ |
| .................................................. | ○ | ○ | ○ | ○ | ○ |
| .................................................. | ○ | ○ | ○ | ○ | ○ |
| .................................................. | ○ | ○ | ○ | ○ | ○ |
| .................................................. | ○ | ○ | ○ | ○ | ○ |
| .................................................. | ○ | ○ | ○ | ○ | ○ |
| .................................................. | ○ | ○ | ○ | ○ | ○ |
| .................................................. | ○ | ○ | ○ | ○ | ○ |

Things that can be done differently next time:

○ _____          ○ _____

○ _____          ○ _____

○ _____          ○ _____

○ _____          ○ _____

# More Powerful Than Gold

Years passed by and the senior commander was slain in a big battle.

Hannibal took over his position and fought many other small battles. Thanks to several victories, he won the trust of the Shophets and senators. They were the highest government powers. Furthermore, he convinced them that Carthage couldn't grow as long as Rome was in its way. He convinced them that an attack would be their best defense.

Finally, the Shophet of the armed forces approved the attack. There was only one problem. Hannibal didn't have enough resources. He needed 100,000 infantrymen for the war. The resources that were provided by the Shophets and senators could barely cover the costs for 20,000 men. He needed fivefold this number of resources to finance infantrymen, weapons, ships and supplies.

Thinking about how to fix the lack of resources, his assistant Maharbal assumed that gold would solve the problem. They just needed to convince the rich people in Carthage to fund the war. Maharbal knew a man named Hasdrubal, who owned half the ships in Carthage and many lands. Not only could he help with his own wealth, but he also had access to a great network of wealthy people who could help too.

Hannibal and Maharbal visited Hasdrubal late at night. He had two strange guests, but they were about to leave anyway. He welcomed them in his royal suite.

"It is a glory to have you men in my house! To what do I owe this honor?" Hasdrubal bid them to take a seat.
"Times are changing, Hasdrubal! Danger lurks at the north side of the sea. It is getting stronger day by day," Maharbal started the conversation.
"Yes, I agree. I am in contact with many senators, it is the concern of every citizen. I hate war, but I know it is sometimes inevitable," Hasdrubal empathized.

"We are here to see how much you are ready to give for this purpose."

Maharbal got down to bases while Hannibal kept silent.
"I am afraid I can't help. I am not interested in wars. All gains made by war are taken by the government," Hasdrubal elegantly denied.
"Well...you are not interested yet! But as soon as Rome attacks Carthage, you will be interested in saving your wealth and your estates," replied Maharbal in a passive-aggressive tone.

"I see that Hannibal is quiet this whole time. Carthage's greatest strategist might know that fear is not the motive of the riches. He might know also that the rich are not afraid of crises. They welcome them to multiply their fortunes. While most people are trying to prevent losses, the rich try to make profits. Putting money into preventing losses is already lost money. It only makes wealth shrink instead of growing it. That makes me wonder why Carthage's greatest commander honors me with his presence?" wondered Hasdrubal.

"You are maybe interested in a position in the senate, or you may know affluent people who want that?" Hannibal broke his silence.

"This sounds like the voice of reason. Being a senator is an attractive offer for any wealthy man. However, my good relations with most of the senators already meets my needs. I can surely name you a few people who would be interested in your offer. However, if you would allow me, I would like to give you some advice instead. It might be worth more than the gold you are seeking," sympathized Hasdrubal.

"We are looking for support. Advice from a wealthy man who knows his onions might be the best support," answered the wise Hannibal.

"You and I are clearly speaking the same language. Unsurprisingly, we both have a lot in common. We have to manage, influence and pay people. If we don't do that well, we would be stabbed in the back at any time. Your battle's costs are getting bigger, faster than your achievements are. I know a Babylonian saying. It implies that gold splits away from those who force it to do what it cannot do. If you want to lay out more money and wait for better results, you better wait

*until the salt blossoms.*

To try many times is advisable, but to keep doing what is not working is insane. It is hard to know when the right time is to stop. Only gut feelings can tell you, and I can bet you have a great one," Hasdrubal sincerely appraised Hannibal. "Obviously, your words carry weight. However, I need to buy slaves. I must pay a lot for supply and infantry. The only thing that can do that is gold. Is there something that buys what gold can't buy?" asked Hannibal.

"There are a lot of things that money can't buy. Let me name a few that are relevant to your case: *hope, trust and faith!* I can pay a man to change his religion, but I can't buy his faith. People believe that there is nothing more precious than someone's own life. But I can assure you that children are more important than that. Furthermore, the only thing that is

*more precious than the son, is the son of the son.*

The descendants are more valuable to men than their own lives. Yet, they sacrifice their newborns to Tanit, the Goddess. Tanit represents trust, hope and faith. Everyone knows Tanit. Does everyone know your mission?

Give your mission a name. Make it hopeful, faithful and trustworthy. Then your mission will be more believable than the Goddess. When you do that, there will be no need for either payment or fear. Hope enables men to do what they can't do. Faith lets them travel to the other side of the sea, even without ships. It may lead elephants to cross snow. Trust convinces men to give more than their gold. It leads them to give away their lives and the lives of their descendants."

"It makes sense," said Hannibal and he kept silent for a moment. He then stood up and said, "Carthage or Rome! We have to leave, Maharbal. We have a mission to spread."

"Great! One thing I would like to add," said Hasdrubal while both men were leaving his house, "Be careful! Men are like gold. They split away from those who force them to do what they aren't able to do. Don't ever be surprised if the

*manliness appears and disappears.* "

Those were Hasdrubal's last words, before both men took their leave.

Not receiving any money from Hasdrubal made Maharbal upset. He left the house saying, "What a grumpy and greedy old man."

Hannibal smirked and said, "We have more now than what we originally came for. He gave us advice, something that is more powerful than gold."

## Some of the hidden lessons:

*~Bidding for money~*

- Even Hannibal was not able to fight Rome alone as a single man.

- A lack of money shall not be your excuse.

- Find the root of the problem before searching for money.

*~Threatening Hasdrubal~*

- Fear cannot be a long-term motive.

- Fear is not the right motive for getting rich.

*~Hope, trust and faith~*

- Having hope, trust and faith is more powerful than money.

- There are many things that money can't buy.

*~Your own lessons~*

- ………………………………………..………………………………..

- ………………………………………..………………………………..

- ………………………………………..………………………………..

# Swords vs. Sickles

"Dear Senators, Hannibal is outside bidding you for an urgent meeting," informed one of the servants to the senate.

"He is back again to plea for more gold. We already offered him the amount for 20,000 infantrymen. We must deny his request," said one of the senators to the rest.

"He said he needs an amount for 100,000 men. We can give him half of what he needs just in condition. We can increase the city tax in return. This is how we can get our money multiplied in the next coming years," said another senator.

Most were greedy enough to like this idea and allowed Hannibal to enter.

As soon as he entered, he started his speech: "Dear Senators! It's wonderful witnessing you stay up at night discussing the issues concerning the Carthaginian citizens. Tonight, I bring you a formula for how you can provide the money to pay for the costs of attacking Rome. I propose that instead of providing me the promised amount of money, you can just give me half of it, with three years of tax reduction."

The senators were impressed. First, they were not obliged to give him more money. Furthermore, they only had to pay half. However, there was another problem.

"Reducing the tax would not be sufficient for the government costs," shouted one of the senators.

"You can still keep the sea trading tax without any reduction," Hannibal tried to convince the senators. He continued, "It brings the most gains anyway. You can also keep all your ships, I don't need them anymore. The most important thing is reducing the tax on all Carthaginians."

All were agreed. The formula was approved by the Shophets too.

"The merciful Hannibal has reduced the tax!" This sentence was heard in every corner of Carthage. It was the first time tax reduction had happened in the era of Carthage. Hannibal became a figure of mercy.

The senators envied Hannibal because he was being increasingly admired by the Carthaginians. To get their appreciation too, the senators engraved the face of Hannibal onto silver coins. But this had an unexpected outcome, and only made Hannibal even more popular. What Hannibal didn't know was that close enemies were shaped secretly in his rows.

Whenever Hannibal won a battle, he came back to Carthage and celebrated his victory. In every corner, men yelled, "Carthage or Rome!" and started to volunteer to join the infantries.

Hannibal sent messengers to surrounding lands and asked for their support. Hannibal's infantry consisted of about 50,000 men. However, their faith on the mission made their strength equivalent to more than 100,000 men. Nevertheless, the more Hannibal's strength and popularity increased, the more betrayers snuck into the senate.

One of Hannibal's messengers was sent to a kingdom on the North-western side of Africa. The kingdom was under the dominance of Carthage and possessed a lot of cavalry. The messenger intended to convince their king, Massinissa to provide some of his cavalry for Carthage's mission.

After greeting the king, the messenger started the discussion and said, "Carthage or Rome! Dear Massinissa, Hannibal not only campaigns for you in the senate to reduce tax, but also seeks to eliminate Carthage's opponent. Once Rome is destroyed, a lot of costs will be avoided. This is how he aims to reduce tax again, even more this time."

"Soon it would be the cereal harvest season. I doubt I can provide any fighter, animal or anything else," rejected the king. The messenger tried to convince him,

*"Wear your sword to wear your sickle."*

But Massinissa insisted, "It is not really our war. We pay tax either to Carthage or to Rome. We better stay safe here."

"Rome doesn't want any tax. They want the demolition of Carthage and its surroundings," answered the messenger.

"As long as they avoid our kingdom, it is not our concern," argued the king.

"Your kingdom will not be an exception. They will destroy it and make its soils infertile." The messenger was wasting his energy.

"As long as they avoid my dwelling, I can tolerate it," said Massinissa.

"The king's house will be destroyed first and they will batter you," replied the messenger.

"I would tell them, just

*avoid my head and hit.*"

The king ended the discussion.

The messenger had never seen anyone so stubborn and left. Hannibal knew that not everyone would believe in his mission. As a result, he simply paid to get a few thousand of their cavalry.

Thanks to Hasdrubal's advice, Hannibal spread hope, trust, and faith among his army. He was able to cross the alps with thousands of infantrymen and invaded the territory of Rome. Meanwhile, Rome counterattacked Carthage, but this was its last hope. This led Hannibal to spare Rome and save Carthage. His assistant Maharbal didn't encourage him to do so, but Hannibal ignored his advice.

During that time, many events happened without Hannibal's knowledge. Senators spread the rumor that Hannibal was hungry for war and power. They found him guilty of losing family members in the war. Moreover, Hannibal didn't know that Massinissa had formed an alliance with the Romans.

Betrayal, ignorance, and patriotism led Hannibal to try and extinguish Rome's last hope. He forgot the power of hope and lost most of his army.

**Some of the hidden lessons:**

*~Massinissa~*

- Some people will never believe in you even after you succeed.

*~The envious senators~*

- The most dangerous betrayers are your confidants.

- Your success will create enemies.

- Don't take success for granted.

*~Your own lessons~*

- .................................................................................

- .................................................................................

- .................................................................................

**Tools for Practice**

This book is just the beginning of many self-development tools. Various apps, games and other tools are in development and will follow. They will all be published at *www.thetyrianpurple.com*

# Closure

Dear reader,

Some may think that these stories are just imagination. However, the harvester of Mactar's story[2g] might change their point of view. Although the stories in this book took place in the ancient age, their lessons are timeless and apply especially nowadays.

Today, Tunisia possesses not only the Carthaginian knowledge, but also their *geographical position, mineral resources, agriculture, human resources* and all the new technologies. Let's review each point and see how they are being misused.

**Geographical position:**
The port* of Carthage was razed to the ground. Close to it, there is a port named Halq al-Wadi, now. The logistics there are disastrous. Vessels wait for months to get their charges unloaded.

**Mineral resources:**
Although Tyrian purple worth much more than gold, it alone cannot raise finances for a whole country. On the other hand, Tunisia has discovered many other resources like oil, gas, solar energy and much more. Still, the Tunisian government is in debt.

**Agriculture:**
Tunisian agriculture is living through its worst times. Original seeds were abandoned over decades. The agricultural sector has been transformed into a big mess. Nevertheless, the country is

* Visit the port of Carthage https://maps.app.goo.gl/FwKtQPewuM43uVuy5

capable of producing wheat, olive oil, dates, oranges and much more.

**Human resources**:

Although Tunisia possesses thousands of people like Ellisa and Hannibal, they have all either left, or are thinking of leaving the country, legally or illegally.

One thing that Carthage possessed, but which Tunisia does not, is a mission. People are living with a lack of perspective. They don't lack leaders like Dido or Hannibal. They just need a mission that gives them

*hope, trust and faith;*

Hopefully, you had a pleasant time reading *The Richest Man in Carthage*.
Please tell your friends about your good experience and share it on social media using *#TheRichestManInCarthage*.

We would really appreciate it, and it will help us make more self-development materials.

# Acknowledgment

Thank you God for your uncountable blessings that I have received.

I would never have been able to write a single word of this book without my illiterate mother's support. Thanking her is beyond the scope of this book.

I would like to thank my father and four sisters for being by my side all the time.

Special thanks go to Manel Bargaoui, my wife, my inspiration, my idol… Words cannot express my gratitude.

# Appendix 1

a. **'Chaos'**

Today this game is known as 'tick-tack-toe'. People used to draw a star and use two kinds of stones, one for X and one for O.

b. **Ten winters of sacrifices**

Despite his losses in the first months, Hasdrubal kept investing in his chicken farming business. Every month he bought a new chicken with his savings. From the eggs, he ate eight and put the rest into brood. Not all the eggs hatched, on average he got 16 chicks out of 20 brooded eggs. We can say that half of the chicks are female and the rest are male. Chicks need about six months to get sexually mature. Female can then produce eggs while the roosters fertilize the chicken. Hasdrubal kept four hens and one rooster. He sold the rest for fodder. Brooding hens don't lay eggs for the whole time, not until their chicks are about 6 months old. Eggs need about 21 days to hatch. For our calculation, we will consider this to be a whole month.

The following tables show how the business developed in just the first year. Losses caused by wild animal attacks, theft, diseases etc. are not taken into consideration. By the twelfth month, the number of chicks and eggs would be hard to count indeed.

| 1. Year of applying the seven laws of gold | | | | | 2. Year of applying the seven laws of gold | | | | |
|------|------|----------|--------|------|------|------|----------|--------|------|
| Time | Hens | Roosters | Chicks | Eggs | Time | Hens | Roosters | Chicks | Eggs |
| Start | 0 | 0 | 0 | 0 | Start | 26 | 6 | 96 | 280 |
| 1. Month | 0 | 0 | 0 | 0 | 13. Month | 35 | 8 | 240 | 352 |
| 2. Month | 0 | 1 | 0 | 0 | 14. Month | 44 | 11 | 384 | 424 |
| 3. Month | 1 | 1 | 0 | 0 | 15. Month | 53 | 13 | 528 | 496 |
| 4. Month | 2 | 1 | 0 | 28 | 16. Month | 62 | 15 | 672 | 568 |
| 5. Month | 3 | 1 | 16 | 28 | 17. Month | 71 | 17 | 816 | 640 |
| 6. Month | 4 | 1 | 32 | 28 | 18. Month | 80 | 20 | 960 | 712 |
| 7. Month | 5 | 1 | 48 | 28 | 19. Month | 161 | 40 | 960 | 784 |
| 8. Month | 6 | 1 | 64 | 28 | 20. Month | 242 | 60 | 960 | 3124 |
| 9. Month | 7 | 1 | 80 | 28 | 21. Month | 323 | 80 | 2256 | 3844 |
| 10. Month | 8 | 1 | 96 | 28 | 22. Month | 404 | 101 | 3552 | 4564 |
| 11. Month | 17 | 4 | 96 | 28 | 23. Month | 485 | 121 | 4848 | 5284 |
| **12. Month** | **26** | **6** | **96** | **280** | **24. Month** | **566** | **141** | **6144** | **6004** |

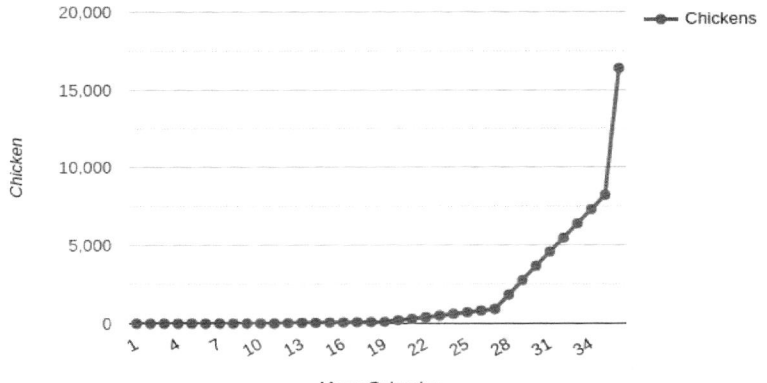

**Hasdrubal's Chicken Husbandry - Chickens**

III

Let's see how Hasdrubal's wealth grew. A hen or a rooster had a value of one piece of silver. Consider that four eggs are equal to one chick. Both have a value of the tenth of a silver piece. The following diagram shows how the wealth would grow in the first two years, in comparison to if Hasdrubal had just saved without investing.

**Hasdrubal's Chicken Husbandry - Wealth**

IV

# Appendix 2

a. Krahmalkov, Charles R. (2000). Phoenician-Punic Dictionary. Leuven: Peeters. ISBN 90-429-0770-3.

b. Ottavo contributo alla storia degli studi classici e del mondo antico Arnaldo Momigliano – 1987

c. The Holy Quran Chapter 81, verses 8-9

d. Exposed in the British Museum. Reference: C.263.

e. The Comprehensive Aramaic Lexicon. cal.huc.edu. Retrieved 2021-01-20.

f. The Richest Man in Babylon, ISBN 978-0451205360.

g. Roman Social History ISB13-978-0415426756

# Appendix 3

All idioms and their meanings:

1. *"Flame rather than shame."*
   *Meaning:* The good reputation is very important.

2. *"You are my liver! You are the light of my eyes!"*
   *Meaning*: I can't live without you. Loosing you would make me cry until I went blind.

3. *"Money is the dirt of the world."*
   *Meaning*: Money is not worth it.

4. *"Put money on the dead's mouth, they will smile."*
   *Meaning*: Even an emotionless person would be happy to get money.

5. *"Money turns people evil."*
   *Meaning*: People can become evil when it comes to money.

6. *"What is yours is yours and what is not is not."*
   *Meaning*: Your earning is your destiny.

7. *"Reached the spring but didn't drink."*
   *Meaning*: Failure, even though it was so close to success.

8. *"... saw, what is written on your brow."*
   *Meaning*: Destiny reveals itself when it has already happened.

9. *"Touch of a master is better than ten of an amateur."*
*Meaning*: Professionals are ten times better (or more) than amateurs.

10. *"Three in the hand is better than twenty-three in the tree."*
*Meaning*: What is there counts more than what is left.

11. *"The greedy spends the night creeping."*
*Meaning*: The greedy one makes a fool of himself.

12. *"Trying to get honey out of wasps."*
*Meaning*: Trying something that will never work.

13. *"Poor partnership leads only to loss."*
*Meaning*: A poor partnership is to be avoided.

14. *"99 ants entered the ant hill."*
*Meaning*: Going out of your way for someone who doesn't appreciate it.

15. *"Heritage goes away and handcraft will stay."*
*Meaning*: The inheritance can be spent while the craft continues to bring in money.

16. *"Get hungry and enjoy honey,*
*bind your wife before starting a quarrel, and*
*wear then tear healthy."*
*Meaning*: The necessary is the most gratifying. Build a strong relationship with a woman before you go through hard times with her. Keep your belongings as long as they are not broken or harmful to your health.

17. *"Revive me today then kill me tomorrow."*
    *Meaning*: Looking for an immediate solution without thinking of the consequences.

18. *"New horses but the chariot is the same."*
    *Meaning*: Outer appearances can change, but the result is still the same.

19. *"Play it crazy to survive."*
    *Meaning*: To behave in such a way that people feel pity for you.

20. *"Take your part from the start."*
    *Meaning*: Be the first and don't waste time.

21. *"Man makes mistakes until he'll be wise."*
    *Meaning*: Despite years of experience, mistakes are still made.

22. *"Little money with little money turns to big money."*
    *Meaning*: Small amounts lead to large amounts.

23. *"Sweat to appreciate the bread."*
    *Meaning*: Money earned by oneself is more valuable than money earned by other people.

24. *"Dig a fountain even with a needle."*
    *Meaning*: Where there is a will, there is a way.

25. *"Beggar and swagger."*
    *Meaning*: Poor but arrogant or being poor but spending too much.

26. *"Man makes mistakes even if he is wise,*
    *man makes mistakes until he dies."*
    *Meaning*: Mistakes will always be made.

27. *"Looking for a cold bread."*
    *Meaning*: Wanting something without being willing to
    do anything for it.

28. *"No vendor, no labor."*
    *Meaning*: Absence and lateness lead to loss of earnings.

29. *"The owner of the soul is its healer."*
    *Meaning*: One knows himself best.

30. *"Do like your neighbor or change your labor."*
    *Meaning*: Follow the crowd or make your own way.

31. *"He who was born one night before, possesses one trick*
    *more."*
    *Meaning*: More experience means more ability.

32. *"Offer and forget suffering or suffer and forget*
    *offering."*
    *Meaning*: Decide without regret afterwards.

33. *"By the first blare he wants his share."*
    *Meaning*: Wanting a reward for the slightest effort.

34. *"Selling wind to the ships."*
    *Meaning*: Being a fraud.

35. *"Wool is sold for its heaviness."*
    *Meaning*: Approach slowly, without being fooled by appearances or first results.

36. *"No prevents troubles."*
    *Meaning*: Saying no keeps one out of trouble.

37. *"Neither loves me nor can he resist me."*
    *Meaning*: Not being treated well but also not being left alone.

38. *"Hanging with the group is a holiday."*
    *Meaning*: With the community, hard times can be overcome.

39. *"Real men earn, and millers yearn."*
    *Meaning*: Some earn and others envy.

40. *"Came from behind and said this is mine."*
    *Meaning*: Falsely claiming to be the owner.

41. *"Gurgles melt marbles."*
    *Meaning*: Countless attempts achieve what is difficult.

42. *"Beat me to cry! Pass me and snitch!"*
    *Meaning*: Do wrong and claim to be the victim

43. *"Son of the mouse becomes a digger."*
    *Meaning*: A child shows the same characteristics as its parent.

44. *"The watcher is a warrior."*
    *Meaning*: He who looks from the outside knows what to do.

45. *"Teamwork makes heaviness less."*
    *Meaning*: Teamwork makes tasks easier.

46. *"Small but always a amazing."*
    *Meaning*: Underestimated but amazing.

47. *"Until the salt blossoms."*
    *Meaning*: Until all eternity.

48. *"More precious than the son, is the son of the son."*
    *Meaning*: Parents love their descendants more than themselves.

49. *"Manliness appears and disappears."*
    *Meaning*: One can become weak.

50. *"Wear your sword to wear your sickle."*
    *Meaning*: Fight to enjoy the peace afterwards.

51. *"Avoid my head and hit."*
    *Meaning*: It is none of my business as long as I am not disadvantaged.